The LifE Project

Long-term Initiatives for Flood-risk Environments

By Baca Architects with BRE and
Fulcrum Consulting, Halcrow Group Ltd, Cyril Sweett and LDA Design

Acknowledgements

The LifE Team

The LifE project is one of six projects funded by Defra's Flood and Coastal Erosion Risk Management - Innovation Fund, and is part of Defra's Making Space for Water programme. The Innovation Fund seeks to improve future delivery of flood and coastal erosion risk management, by promoting innovative approaches that contribute towards the development of more holistic and sustainable policy making in the future.

This project was led by Baca Architects and BRE in collaboration with Fulcrum Consulting, Halcrow Group Ltd, Cyril Sweett, LDA Design and support from Dura Vermeer and the Max Lock Centre, University of Westminster.

The Steering Group

John Blanksby, University of Sheffield
Tom Lloyd-Smith, Housing Corporation
Douglas Barnett, AXA Insurance
Jonathan Chapman, Environment Agency
Anna Penning-Rowsell, CABE
Philippa Bloomfield, English Partnerships

Thanks are given to the London Borough of Sutton, Opportunity Peterborough, West Sussex County Council, the Environment Agency and Martin Roberts of Defra for their support.

Publisher

Published by IHS BRE Press on behalf of the LifE project.
IHS BRE Press, Willoughby Road, Bracknell RG12 8FB
brepress@ihs.com

EP 98

ISBN 978-1-84806-101-9

First published July 2009

Authors

The LifE report and LifE handbook were written by:

Robert Barker, Baca Architects
Richard Coutts, Baca Architects
Tia Randall, Baca Architects
Craig Allen, Baca Architects
Vivianne Blanco, Baca Architects
Nick Corker, BRE
Stephen Garvin, BRE
Tessa Parnell, Fulcrum Consulting
David Altabev, Fulcrum Consulting
Marcus Phillips, Halcrow Group
John Palmer, Halcrow Group
Simon Harris, Cyril Sweett
James Wanless, Cyril Sweett
Angus Jeffrey, LDA Design
Chris Laine, LDA Design
Ripin Kalra, Max Lock Centre

with thanks to Jessica Lawrence

For further information go to - www.lifeproject.info

Contents

Foreword

Flood risk in the built environment needs to be managed. We cannot avoid the issues stemming from changing climate and expanding urban environments. It is important then to have a range of clear, practical and helpful tools. This report offers substantial guidance for those involved in achieving flood risk management and sustainable development in new construction works. Its contents underpin the LifE Handbook and as such represent an overview of the technical detail and working methodology resulting from the LifE project.

Every professional discipline (landscape architecture, quantity surveying, services engineering, and so on) brings its own priorities to any development project. Each has its own language, modus operandi and culture. One characteristic of the 'LifE approach' has been to take these independent sectoral priorities and to identify what levels of integration can take place between disciplines. It identifies methods and tools from the professions that are mutually reinforcing when applied at the development scale and therefore offer better value in the long term.

For example, can we specify a green infrastructure that also provides fibre for biomass heat, as well as acting as an amenity space and a 'sponge' for mopping up surplus water and ameliorating the urban heat island? Is it possible to identify where flood resilience needs to be part of a building's specification? Should community buildings be so designed and located to act as local refuges if a flood occurs? Where can investment be best focused for the most benefit for the majority of the population now and in the future?

For many working in the development industry, understanding and conveying information about these issues is difficult even for the technical specialist. The LifE Report illustrates the issues and indicates potential solutions. It considers the background context to flood risk management in the UK, sets out the LifE methodology, illustrates the approach through three case studies, addresses the potential solutions and sets out transferable ideas that can be used throughout the country.

The project included consultation with key stakeholders in its preparation, both at steering group level and through informed comment from statutory agencies. As such I believe that it should be of significant interest to a wide range of the development industry, local authorities and government departments. It is important to stress that the LifE project is not at odds with existing legislation, regulation and guidance but seeks to build on this foundation and to illustrate how developments can come forward in an holistic and integrated way.

Sir Neville Simms

Chairman, BRE Trust

June 2009

The LifE Team

Baca is an award-winning research-led design practice specialising in integrated environmental and high quality design. The practice has a diverse array of projects that includes, one-off houses, social housing, master planning, memorials and commercial projects.

Baca is internationally recognised for innovation in flood resilient and adaptable architecture and spatial planning. The LifE project explores innovative ways to tackle flooding and climate change through the built environment.

www.baca.uk.com

The BRE Trust companies, BRE and BRE Global, are world-leading research, consultancy, training, testing and certification organisations delivering sustainability and innovation across the built environment and beyond. With around 650 research scientists, engineers, architects, surveyors, psychologists, administrators and managers, BRE remains one of the main independent research and consultancy businesses in the UK.

Our mission is to 'Build a better world'. We help our clients create better buildings and communities and solve problems with confidence.

www.bre.co.uk

For over 130 years Halcrow has been a leading international multi-disciplinary consultant, addressing the following sectors: consulting and environment, maritime and coastal, property, transportation, water and power. Halcrow's vision is "sustaining and improving the quality of people's lives". With this vision at the heart of our business, Halcrow seeks to achieve our business objectives in a manner that prevents pollution, causes least harm to the environment and, as far as is reasonably practicable, achieves environmental benefits.

www.halcrow.com

Cyril Sweett is a leading international construction and property consultancy offering expertise in cost consultancy, project management, building surveying and a comprehensive range of specialist consulting services. The group was founded in 1928 and currently has 29 offices throughout the United Kingdom, Australia, France, India, Ireland, Singapore, Spain and the United Arab Emirates.

www.cyrilsweett.com

Fulcrum Consulting is a leading international firm of multi-disciplinary consulting engineers. We have developed leading expertise over the last 25 years in the design of low energy buildings and sustainable communities and infrastructure.

Our objective is to facilitate the ongoing development of the built environment in a way that is compatible with comfort, cost efficiency and low carbon emissions. Our approach to sustainability is through the application of integrated engineering design to buildings and infrastructure to determine the optimum solution for each of our clients' particular needs.

www.fulcrumfirst.com

LDA Design brings together the skills of landscape architecture, urban design, architecture, ecological consultancy, environmental planning and community asset management.

Founded in 1979, LDA Design is a renowned independent design and environment business. Driven by a commitment to shape the world for the better, we achieve bespoke creative solutions through the ideas-led culture we have built and the skill and experience of our team. We help our clients regenerate communities, create special places, manage resources and realise their development and commercial goals.

www.lda-design.co.uk

Executive Summary

The UK government has advocated an integrated approach to flood and coastal erosion risk management in order to reduce environmental impact and improve sustainability. When combined with the objectives for creating sustainable communities, this poses questions of our built environment, such as: *'What is an integrated approach? What needs to be integrated and why? What will it look like?* and *How much will it cost?'*

The LifE (Long-term Initiatives in Flood-risk Environments) project is a major research project that seeks to answer these questions and aims to create a more integrated, sustainable and responsible approach to development and flood risk management.

Planned over five stages, the objective is to develop clear guidance for development, demonstrate this in practice through built pilot projects, and establish a national interactive LifE toolkit for use by local authorities and planners. This report, and accompanying handbook, form the first stage of the project to establish generic principles for integrated design on sites around the UK.

This LifE report documents the findings of the first stage of the project and conclusions formed from three conceptual masterplans, developed as case studies and examples of integrated development.

The three conceptual masterplans for developments of approximately 2000 new homes on each site are described and illustrated in detail within the appendices. These sites are located in:

- Hackbridge within the upper catchment of the River Wandle,
- Peterborough within the middle catchment of the River Nene,
- Littlehampton within the lower catchment of the River Arun.

Proposals for mixed use developments were used as case studies for the project.

Generic concepts were distilled from the case studies and organised and presented by location.

- The LifE proposals used non-defensive means of managing flood-risk, to make space for water, with floodwater able to pass over and into parts of the development site in a controlled manner.
- Developments were designed to Zero Carbon, with energy requirements met by renewable power generated on-site.
- Houses were designed to Code for Sustainable Homes Level 6 and non-domestic buildings designed to BREEAM Excellent.
- Sites were planned to replicate greenfield run off rates for rainwater and to provide areas for outdoor play set by the local planning guidelines.

The project demonstrated the difficulty in achieving these environmental standards and the premium in cost required to exceed current (2009) standards. It showed that it was possible to achieve low to medium density zero carbon development in less urbanised areas. It also showed that several forms of renewable energy generation, in particular wind, are needed to supplement the use of biomass to provide sufficient power.

The cost of managing flood risk was shown to represent only a very small proportion (1% to 9%) of the total capital cost of the development and the cost of providing renewable power represented the largest increase, rising from 8% to 24% of the total capital cost.

In each case study the means of managing flood risk provided wider benefits to the development through creating space for renewable energy and space for recreation, within walking distance of properties.

Climate change was shown to increase the flood risk to each site, but it was found possible to manage the increased level of risk with design measures introduced at the time of construction.

The location of the site within the river catchment was found to have a bearing on flood risk. The site located in the upper catchment had the least flood risk and the site in the lower catchment had the greatest. The most important flood risk management measure within each case study site (after avoidance) was distilled into the following generic messages:

- Upper catchment - Let rain slow (rainwater attenuation)
- Middle catchment - Let rivers flow (flows and drainage channels)
- Lower catchment - Let tides go (flood diversion and avoidance)

The catchment location was found to be an aid to high level assessment of flood risk. However, each site was found to have unique issues and problems which needed to be assessed by an expert team.

The LifE project found that integrated planning can help to achieve zero carbon development more cost effectively, it can create better quality environments to live and work in, it can help manage flood risk and it can help adapt to climate change in the future.

In some cases the LifE approach may help to unlock sites that are currently not considered suitable for development by managing and reducing flood risk overall.

The LifE Handbook

The accompanying LifE handbook provides an introduction to flooding and development. It is to be used by decision makers, designers and developers before and during the early stages of design, in conjunction with policy documents, such as Planning Policy Statement 25 (PPS25).

The handbook is available from www.brebookshop.com
see page 200 for further information.

Introduction >

Introduction

The aim of the LifE project was to identify ways in which the means of managing flood risk could be integrated within development sites, to help meet objectives for zero carbon, high quality, sustainable development at minimal cost.

As the population and its expectations have grown, the demand for housing, employment, energy and food have risen. The growing expectations have required intensification of yields, outputs and housing density within a finite land. The growing demands have resulted in rising carbon dioxide emissions, fuelling climate change and global warming.

Around 5 million people, in 2 million properties, live in flood risk areas in England and Wales[1], and 158,000 homes are at risk of flooding in Scotland[2]. Thousands of new homes are currently planned within UK floodplains. National policy has set a target for at least 60% of new housing to be located on previously developed (brownfield) land. Some brownfield land is at risk of flooding.

Climate change (greatly accelerated by man's activities) could result in land being lost to rising sea levels or subject to more frequent and extensive flooding. Housing growth could result in rising carbon dioxide emissions, exacerbating climate change.

New building developments will need to reduce their impact on the environment, if they are to form part of the solution and not the problem (figure 1). Yet they will also need to cope with and manage the changes in their environment, including the possibility of increased frequency and severity of flooding.

The LifE project seeks to address the need for new homes, increased risk of flooding and heightened environmental standards simultaneously, by integrating three approaches outlined below and illustrated in figure 2.

Living with Water	Adapting to increased flood frequency and severity, likely to happen with climate change
Making Space for Water	Working with natural processes to provide room for the river and sea to expand in times of flood and reduce reliance on defences, where possible.
Zero Carbon	Providing all energy needs from renewable resources on site, such as wind, tidal and solar power.

These approaches were adopted in combination with Code for Sustainable Homes Level 6[3], BREEAM excellent, greenfield run off rates, Lifetime Homes[4], National Playing Fields Association Six Acre Standard[5], Secured by Design[6], Water Framework Directive[7] and other current (2009) design standards.

The project identified ways in which flood risk management may be integrated into sustainable, zero carbon, responsible development.

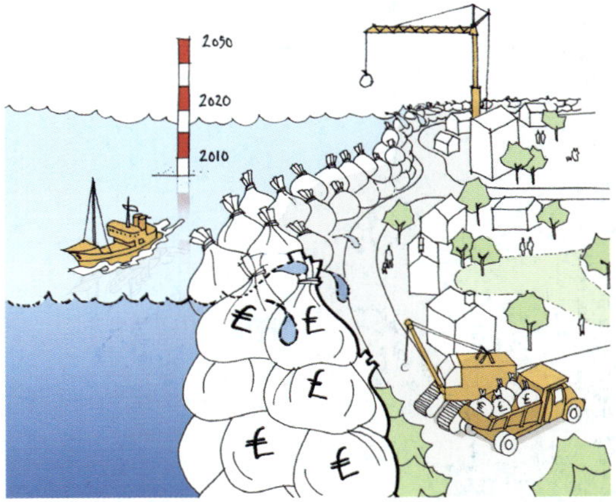

Figure 1 - This is not the only solution

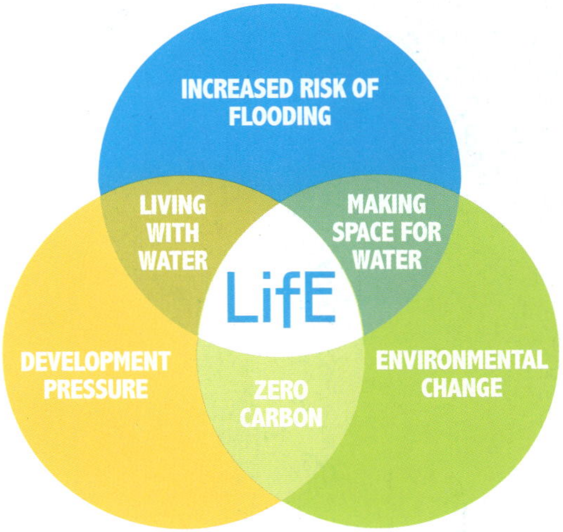

Figure 2 - The LifE approach

The LifE Approach and Context >

LifE Approach

LifE Approach

The LifE approach is an integrated design approach, that uses long-term and adaptable non-defensive flood risk management measures, which provide wider benefits to the development and community; and therefore reduce the economic obstacles to delivering responsible and sustainable development.

The means of managing flood risk explored within the conceptual proposals were designed to have alternative uses for the majority of the time that they were not required to cope with flooding.

Integrated Design

An integrated design approach seeks to use land and water assets for multiple functions, such as selective development, recreation, renewable energy production, local food production, water storage and flood alleviation. Figure 4 illustrates how an integrated design approach combines the water cycle strategy (rainwater attenuation, flood storage, waste water treatment) with the renewable energy supply network (biomass, wind power, centralised energy) with development in flood risk areas (locating more vulnerable uses in less vulnerable areas and use of flood resilient buildings).

Non-Defensive Flood Risk Management

A non-defensive approach to flood risk management that works with natural processes can help to reduce the ongoing maintenance costs associated with defences, reduce the residual risk and increase awareness.

Instead of keeping water out, developments are designed to allow flood water and rainwater into, over or around the sites in a controlled and predetermined manner. The intention is to create more adaptable and intuitive landscapes, improving awareness of flood risk and potentially reducing risk to other areas (figure 3 illustrates the potential implication of a defensive system).

Long-term Initiatives

A development that intrinsically provides flood resilience, through good design and planning, should give insurers and financiers the confidence to provide affordable, long-term policies and investment.

A versatile and adaptable flood risk management strategy is required, which through an explicit response to specific risk and vulnerability, can cope with the increasing unpredictability and severity of flooding resulting from climate change. Secondary and tertiary mechanisms to cope with flooding need to be integrated into design and planning so that protection is not just reliant on a single line of defence.

Furthermore, the concept of continuity of daily life, where possible, before, during and after a flood, should help to avoid the detrimental economic and social impacts that can result from flooding.

Wider Benefits

Flood risk management measures can be used to provide wider benefits for the development and the community such as space for outdoor recreation, space for renewable energy, housing and job provision and place making.

Identifying the potential to create wider benefits at all scales of design can help to improve the quality, success and sustainability of developments.

Reducing Economic Obstacles

Using available resources efficiently and combining functions may reduce implementation costs.

Responsible Development

This should encourage greater environmental improvements to be achieved, higher quality design and more sustainable developments to be built.

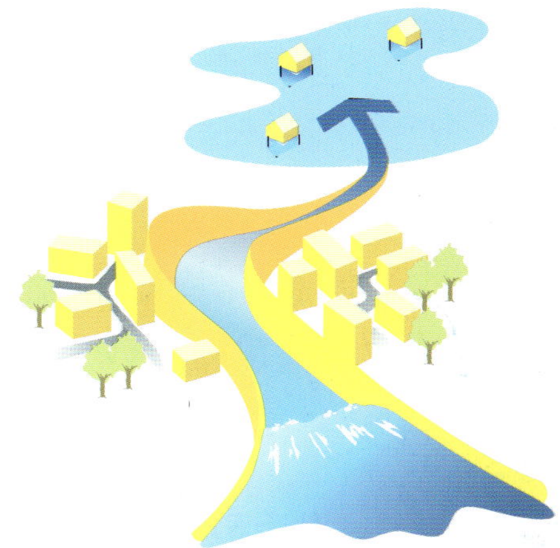

Figure 3 - Consideration of the wider context is important to avoid adversely affecting neighbours

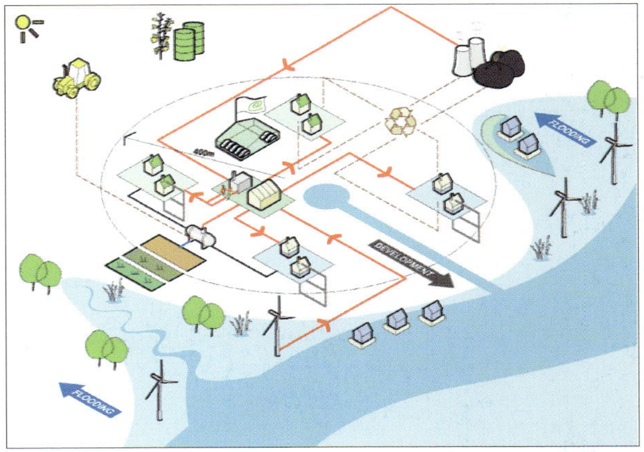

Figure 4 - Diagram of an integrated spatial plan, showing selective development integrated with flood risk management measures and sustainable development

Context > Flood Risk Management

This section briefly describes some recent national and international initiatives and programmes, providing context for the LifE project.

Making Space for Water

National policy responsibility with respect to flood and coastal erosion risk management in England is developed and set by Defra.*

Defra's policy is to reduce risk to people, property and the environment from flooding and coastal erosion through the provision of defences, flood forecasting and warning systems, increased flood resilience of property, beneficial land management changes and discouragement of inappropriate development in areas at risk of flooding.

Making Space for Water[8] is the cross government programme taking forward the developing strategy for flood and coastal erosion risk management in England. The aims are 'to manage the risks from flooding and coastal erosion by employing an integrated portfolio of approaches which reflect both national and local priorities, so as to:

— reduce the threat to people and their property;
— deliver the greatest environmental, social and economic benefit, consistent with the Government's sustainable development principles;
— secure efficient and reliable funding mechanisms that deliver the levels of investment required to achieve the vision of this strategy.'[8]

* The Scottish Government has overall responsibility for policy on flood risk management in Scotland. However, the primary responsibility for protecting land from flooding lies with the owner of the land concerned. A function of the Scottish Environment Protection Agency (SEPA) is to assess flood risk. SEPA published maps on the basis of limited information on their website. Flood risk policy in Wales is set by The Welsh Assembly Government (WAG). Flood risk policy in Northern Ireland is set by The Department of Agriculture and Rural Development NI (DARDNI)

Making Space for Water - Vision

The vision of Defra's Making Space for Water (MSW) strategy is to allow space for water so that we can manage the adverse consequences for people and the economy that can result from flooding and coastal erosion while achieving environmental and social benefits in line with wider Government objectives.

The LifE project is funded under the MSW programme which consists of 25 projects within four broad themes:

— A holistic approach to managing flood and coastal erosion risk;
— Achieving sustainable development;
— Increasing resilience to flooding;
— Funding.

Making Space for Water - Key Messages

The MSW strategy is promoting a number of messages, of which the most relevant to the LifE project are:

— A holistic approach that embeds sustainable development in all actions and decision-making related to flood and coastal erosion risk management;
— Working with natural processes where possible to increase sustainability, reduce costs and provide environmental benefits, including for water quality and biodiversity;
— Promoting adaptation to a changing coast and fluvial flooding situation as it would be unsustainable to maintain all current defences in the face of likely impacts of climate change;
— Promoting positive parallel outcomes of flood risk management schemes: restoration of habitat and ecological benefits, recreational and regeneration opportunities with multiple sources of funding where possible.[8]

Context > Flood Risk Management

Making Space for Water - The Innovation Fund

The Innovation fund project was established to fund new projects that:

- Demonstrate innovative solutions to deliver flood risk management in ways that maximise the environmental, social and economic benefits of the flood risk management programme; and,
- Deliver both tangible flood risk management benefits and wider benefits. The project will consider, in the light of the projects undertaken, whether there are any changes to policy, funding or delivery that would encourage delivery of sustainable flood risk management.

Making Space for Water - Flood Risk Assessment

Since the inception of the Making Space for Water programme the government has developed a hierarchical flood risk assessment and management structure. The aim is to provide a holistic approach to flood management based on examining whole river catchment systems and encouraging collaboration across political boundaries. This includes:

- Regional Flood Risk Appraisals (RFRA)
- Catchment Flood Management Plans (CFMP) and
- Shoreline Management Plans (SMP)
- Strategic Flood Risk Assessments (SFRA), which indicate flood zones and appropriate sites for development

The Environment Agency has committed to completing a Catchment Flood Management Plan for every major river in England and Wales.

Making Space for Water - PPS25

Planning Policy Statement 25[9], which replaced PPG25 in December 2006, provides policy guidance for Development and Flood Risk.

The LifE project work has examined the application of PPS25 and other Planning Policy on three study sites to determine sustainable ideas that might be relevant to other sites at risk of flooding and those that aim to deliver zero carbon developments.

Making Space for Water - The LifE Project

The LifE project work is relevant to sites being considered for development in areas at risk of flooding, particularly from the river, and that can be demonstrated to pass the 'Sequential Test'. Figure 5 shows the LifE project in the context of national responsibilities, based on figure 2.2 from the PPS25 practice guide.[10]

Pitt Review - Learning Lessons from the 2007 Floods

The Pitt Review[11] examined the floods of 2007 and set out recommendations for future changes in flood risk management. The most relevant messages to the LifE project were:

- The call for 'urgent and fundamental changes in the way the country is adapting to the likelihood of more frequent and intense periods of heavy rainfall'.
- 'We...ask councils to strengthen their technical capability in order to take the lead on local flood risk management. More can be done to protect communities through robust building and planning controls.'
- 'Many submissions to the review call for a complete end to building on the floodplain. This is not realistic'[11]

Figure 5. Spatial Planning Process and the stages influenced by the LifE project

Population Growth

The United Nations predicts that the global population will rise from 6 billion in 2000 to between 7.5 billion and 10.6 billion by 2050.[12] The population is expected to continue to rise towards the end of the century. Rising standards of living in developing countries, particularly India and China, are likely to result in an increase in carbon dioxide emissions.

Housing Demand and Growth Areas

The Deputy Prime Minister launched the Communities Plan for England (Sustainable Communities: Building for the Future)[13] in February 2003. The plan sets out a long-term programme of action for delivering sustainable communities in both urban and rural areas. It aims to tackle housing supply issues in the South East, low demand in other parts of the country, and the quality of our public spaces. Following Kate Barker's 2004 report into housing affordability the government established an aim to increase housing supply in England to 200,000 per year by 2016.[14]

The Plan includes not just a significant increase in resources and major reforms of housing and planning, but a new approach to how we build and what we build.[13]

The Communities Plan outlines a £38 billion programme of action for economic, social and environmental development in both urban and rural areas in England. Four areas, shown in figure 6, were identified to play a key role as national priorities for regeneration and growth:

- Thames Gateway,
- Ashford in Kent,
- London-Stansted-Cambridge-Peterborough,
- Milton Keynes/South Midlands.

Figure 6 - The communities plan growth areas

Sustainable Communities

A sustainable community is considered to embody the principles of sustainable development, defined as:

- Balancing and integrating the social, economic and environmental components of their community,
- Meeting the needs of existing and future generations,
- Respecting the needs of other communities in the wider region or internationally to make their own communities sustainable.[13]

The Code for Sustainable Homes

The Code[3] is intended as a national standard to guide industry in the design and construction of sustainable homes. The Code measures the sustainability of a home against nine design categories: energy, pollution, water, health and well-being, materials, management, surface water run off, ecology and waste.

The Code introduces minimum standards for energy and water efficiency at every level of the Code, requiring high levels of performance in these areas for achievement of a high Code rating. Sustainability is measured as a complete package and graded in Code levels with Level 6 being the highest attainable. To achieve Code Level 6, homes must be designed to be zero carbon.

Zero Carbon

In response to the Stern Review[15], the government established a target date of 2016 for all new homes to be built to zero carbon.[14]

A zero carbon building is one that uses renewable energy, produced on the same site as the building, to meet all of the energy demands in operation. The LifE project applied this approach to all buildings and infrastructure associated with the conceptual developments. The definition used for the project is:

> *All building energy demands are met by the use of low or zero carbon energy technologies, by on-site energy generation. The means of energy generation should draw on the resources of the development site where possible. Where it is not physically possible to meet all demands on-site, use of off-site electricity generation is allowed, provided it is accredited renewable generation capacity outside of the Renewables Obligation (RO). However, it must first be demonstrated that all viable on-site generation opportunities have been exhausted.**

* The methodology used to define 'zero carbon' differs from the current (2008) methodology set out in the Code. See Appendix D for further explanation.

The Intergovernmental Panel on Climate Change

In 2007 the Intergovernmental Panel on Climate Change (IPCC) concluded that:

'Most of the global average warming over the past 50 years is very likely due to anthropogenic (man-made) greenhouse gas increases'.[16]

It also concluded that global warming (caused by greenhouse gas increases) had already resulted in sea level rises and would continue to do so for centuries to come, even if greenhouse gas concentrations were stabilised.*

The IPCC model results indicated that sea level rise could be 0.59 m by the end of the 21st century under the A1FI scenario. However, it also states that:

'The sea level projections do not include uncertainties in climate-carbon cycle feedbacks nor do they include the full effects of changes in ice sheet flow, because a basis in published literature is lacking. Therefore the upper values of the ranges given are not to be considered upper bounds for sea level rise'.[16]

Sea level rise and human development are together contributing to losses of coastal wetlands and mangroves and increasing damage from coastal flooding in many areas.[16]

The Flood and Coastal Defence Project Appraisal Guidance FCDPAG3, Supplementary Note to Operating Authorities – Climate Change Impacts, Defra, October 2006[17] identifies sea level rise predictions for various regions around the UK which have been adopted by PPS25[9]. This guidance is to be reviewed on an ongoing basis and building on the work of other key research areas, notably from the Intergovernmental Panel on Climate Change and from UK Climate Impacts Programme (UKCIP).

Kyoto Protocol

The Kyoto Protocol[18], ratified in 1997, sets legally binding emissions reduction targets on the developed countries that ratified it.

Under the Kyoto Protocol, the European Union and its Member States have agreed to meet a joint target of 'an 8% reduction in greenhouse gas emissions below 1990 levels by 2012'. Under the EU bubble agreement the UK agreed to reduce its emissions to 12.5% below 1990 levels by 2012. The UK is expected to meet this target.

The Kyoto protocol ends on December 31st 2012. The UK and other EU nations are calling for more demanding reductions, in the order of a 60% cut in emissions from 1990 levels by 2050, to be imposed in a successor to the Kyoto Protocol.

The UK Climate Impacts Programme

Since 1997 UKCIP has been working with the public, private and voluntary sectors to assess impacts of climate change.

The package of climate change scenarios for the UK has the full title of 'UK 21st Century Climate Scenarios' known as UKCIP09. This replaces UKCIP02.[19]

It is not known how these will affect the current flood risk management strategies but it is unlikely that any relaxation of regulatory requirements will occur.

* Anthropogenic warming and sea level rise would continue for centuries due to the time scales associated with climate processes and feedbacks, even if GHG concentrations were to be stabilised.[16]

Methodology >

LifE Project Process

This section outlines the process used to develop the LifE approach and explains in more detail the key stages of this process that might also applicable to other sites.

The LifE Project Process

The LifE project was divided into six stages, shown in figure 7. Work was carried out in a structured but flexible manner to allow innovation and conceptual ideas to be incorporated. The process is described below:

- A set of guiding principles was developed against seven planning requirements (transport, construction, resources, adaptation, longevity, flood risk management, and land use).
- In parallel, three study sites were selected from entries submitted by local authorities and county councils.
- A set of tools and technologies that might be used for a development were then identified.
- These principles and tools were then explored in conjunction with standard development requirements on each of the study sites to develop a holistic LifE solution.
- An iterative design process, involved developing and eliminating design options through the use of drawings, calculations, analysis and review.
- The designs for each study site were then assessed against policy and industry guidelines, assessed for general sensibility and revised accordingly.
- The most replicable ideas were identified from this work. These are presented in this report and the LifE handbook.[20]
- A steering group consisting of industry representatives reviewed the LifE project at key stages.
- Additional reviews of the case studies were provided by the local site representatives

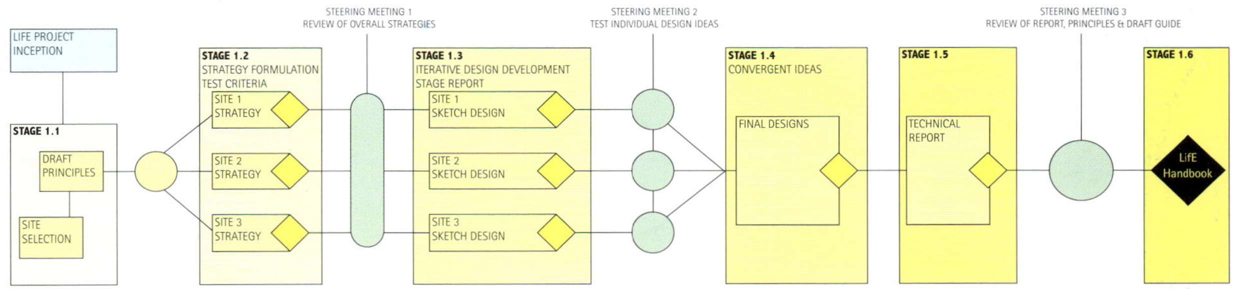

Figure 7 - The LifE project process diagram

The LifE Team and Responsibilities

- Building Research Establishment - contracts management and quality assurance
- Baca Architects - intellectual design lead, project team management, and responsible for the production of all outputs, reports and the handbook
- Cyril Sweett - cost analysis
- Fulcrum Consulting - energy and waste requirements
- Halcrow Group Ltd - hydrological analysis and flood and coastal erosion risk management advice
- LDA Design - landscape and ecology

Additional support was provided during the early stages of the project by the Max Lock Centre, University of Westminster and for the costing of resilient construction technologies by the Dura Vermeer Group, Netherlands.

The Steering Group

The project was presented to the steering group at three one-day workshops. The steering group's responsibilities were to review the project at key stages and give guidance on future stages and outputs. Their comments were integrated into the final outputs. Members of the steering group represented the following organisations:

- The Environment Agency
- English Partnerships
- The Housing Corporation
- Commission for Architecture and the Built Environment (CABE)
- AXA Insurance
- Sheffield University.

LifE Project > Case Study Design Process

Three conceptual masterplans, to form the case studies, were developed following the methodology set out below:

Site Analysis

Information on each site was gathered from a range of sources to determine:

- location and character
- planning policy
- flood risk and flood zones
- river characteristics
- climate change implications
- available resources.

Quantified Brief

The site analysis was used to establish a detailed schedule for each site that quantified a range of development requirements as follows:

- accommodation mix for domestic and non-domestic buildings
- car parking requirement
- area of roads and pavements required
- energy and water demand (based on the accommodation and occupancy schedule)
- public and private outdoor amenity requirement
- surface water drainage required for a 1 in 100 year rain storm
- floodplain compensation required.

Development Objectives

Site-specific development objectives were established in combination with the project-wide objectives to:

- provide 100% energy demand from on-site renewable sources
- reduce car dependency by creating walkable developments (see figure 8)
- reduce use of carbon intensive construction practices
- reduce reliance on flood defences
- use non-defensive flood risk management measures
- create adaptable plans for future climate change
- provide surface water run off rates equivalent to greenfield
- use land for multiple purposes.

Iterative Design Process

An iterative design process, involving exploring and refining options, was carried out for each site to determine the best solution.

These were reviewed by the local authorities and the Environment Agency.

The capital cost of the designs was calculated and the designs further refined.

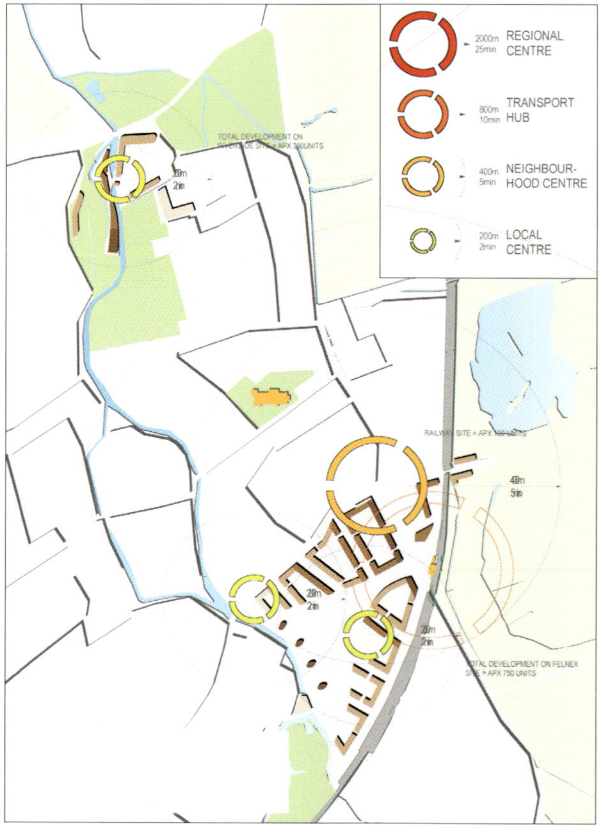

Figure 8 - Site 1 planning based on walking distances

Design Review

The conceptual masterplans were reviewed against statutory, regulatory, and guidance documents; the most relevant of these documents are indicated in table 1.

The case studies are presented in

Appendix A, Site 1 > River Wandle, Hackbridge

Appendix B, Site 2 > River Nene, Peterborough

Appendix C, Site 3 > River Arun, Littlehampton

Accessible Natural Green Space Standards in Towns and Cities: A Review and Toolkit for Implementation[21]	✓
Building Regulations Approved Documents[22] (particularly Part B and M)	✓
Code for Sustainable Homes[3] (Level 6)	✓
Catchment Flood Management Plans[23, 24, 25]	✓
Flood and Coastal Defence Project Appraisal Guidance[17]	✓
European Birds and Habitats Directives[26]	✓
Housing Quality Indicators[27]	✓
Lifetime Homes[4]	✓
Planning Policy Statements (PPS1[28], PPS3[29], PPS9[30], PPS22[31], PPS25[9])	✓
Planning Policy Guidance 17[32]	✓
Raising the Standard: The Green Flag Award Guidance Manual[33]	✓
Secured by Design[6]	✓
Strategic Flood Risk Assessments[34, 35] (emerging)	✓
The Six Acre Standard: Minimum Standards for Outdoor Playing Space[5]	✓
Water Framework Directive[7]	✓

Table 1 - Review documents

Case Studies >

The LifE Case Study Sites

An invitation was sent out to local authorities and regional development agencies around the country asking for major current and future development sites to be submitted for the LifE project study. From those submitted three study sites were selected on the basis of:

- demonstrating a range of flood risk issues and therefore providing an opportunity to explore a range of flood risk management options,
- providing sufficient development scale to explore a wide range of renewable energy technologies,
- containing characteristics and issues likely to occur on other sites around the UK.

The three study sites are located within three distinct river locations and three river catchment locations. Their locations within the UK are indicated in figure 9.

On each site, conceptual proposals for large-scale masterplans of between 1000 and 2000 new homes were considered. These plans were based on the premise of meeting current housing demand, future housing demand and in the case of site 3 financing long-term flood risk management proposals.

The following case studies explored the application of the LifE approach on these sites, using real data provided by the local authorities and the Environment Agency.

Conceptual design proposals were reviewed by local stakeholders, including local Environment Agency officers. These studies do not represent actual proposals being brought forward and any similarity is coincidental.

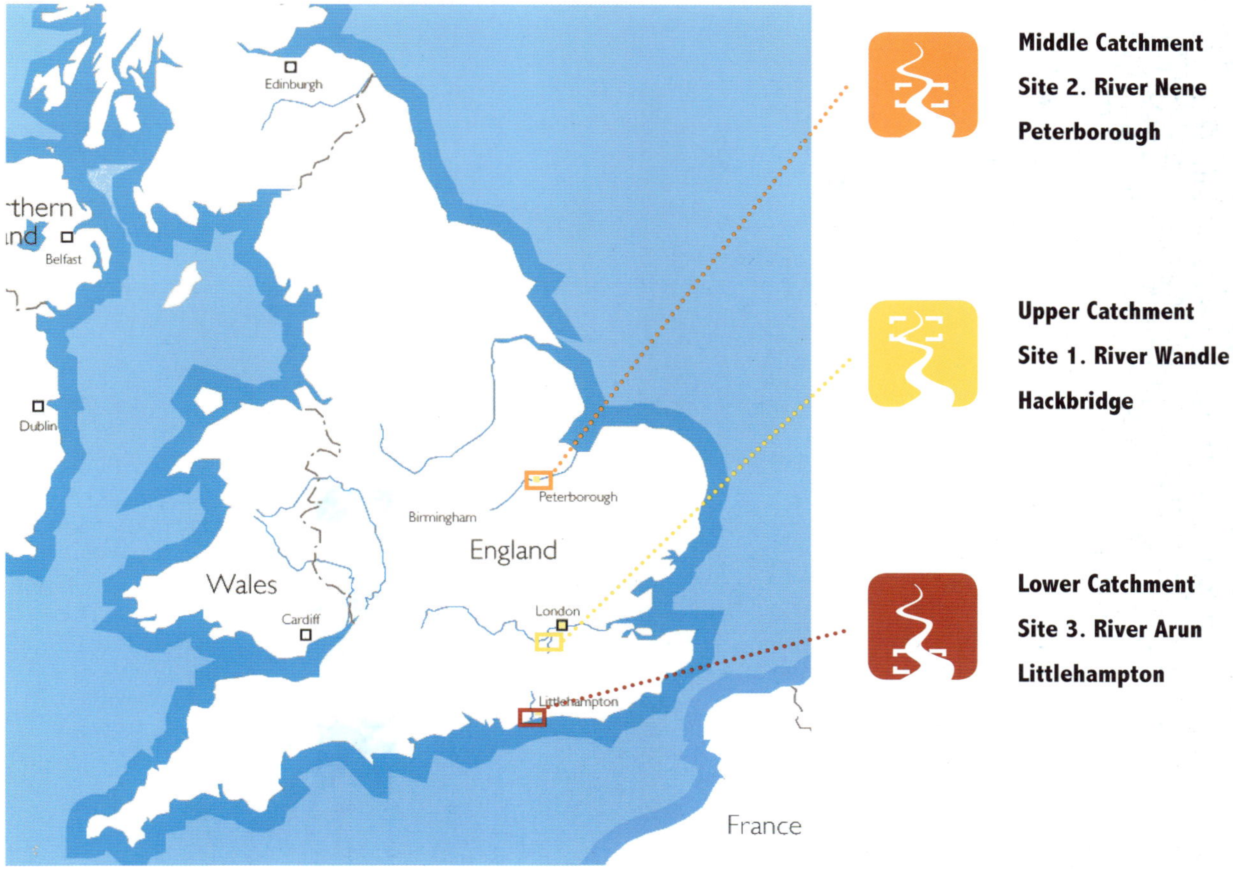

Figure 9 - LifE study site locations

The LifE Case Study Sites

Photographs and flood maps for each site are shown in figures 10 to 15.

Site 1 comprised four sites, occupying a total of 24 hectares, located in the upper catchment of the River Wandle at Hackbridge, South London.

Figure 10 - The River Wandle at Hackbridge

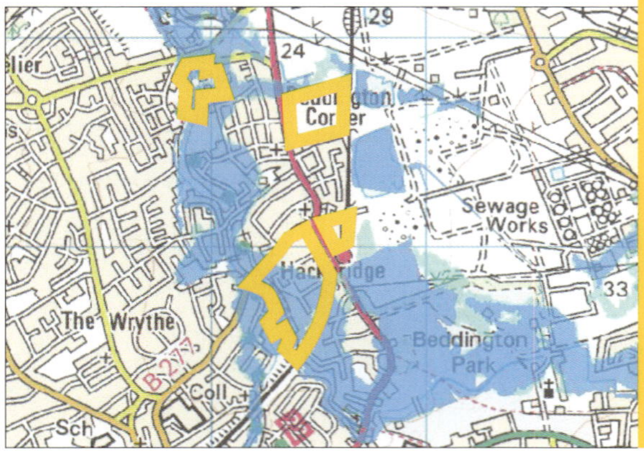

Figure 11 - Environment Agency flood risk map for site 1

Site 2 comprised sites on both the north and south bank of the river, occupying 50 hectares, located in the middle catchment of the River Nene at Peterborough.

Figure 12 - The River Nene at Peterborough

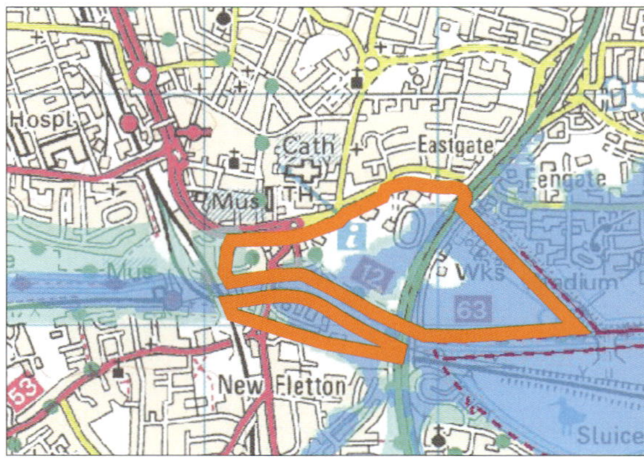

Figure 13 - Environment Agency flood risk map for site 2

Site 3 comprised two sites, occupying 56 hectares, located in the lower catchment of the River Arun at Littlehampton.

Figure 14 - The River Arun at Littlehampton

Figure 15 - Environment Agency flood risk map for site 3

SITE 1 > The River Wandle at Hackbridge

Hackbridge is located in the London Borough of Sutton on the upper catchment of the River Wandle. Flood risk is predominantly from the river, which is flashy and responsive to intense rainfall.

Climate change could result in longer or more intense rainstorms increasing flash floods from overland run off, sewers and the river. Flood Zone 3 is expected to extend from the river's edge further into the site.

A central amenity space, termed the village blue/green (shown in figure 17), could bring multiple benefits to this suburb, such as space for recreation, flood storage and a much needed focal point to the local area,

Higher density development is located in Flood Zone 1 near the railway station. Pedestrian and cycle links would provide wider access to the river; car parking is limited to one space per three homes and a central car club provided.

Mixed residential blocks would look onto communal 'rain gardens' (shown in figure 16) as part of a sustainable drainage system (SUDS). Each garden would provide multiple uses, high quality planting treating surface run off and providing a buffer to a toddlers' play area at the centre, this would be raised above a communal rain-water harvesting system. Pathways, benches and small private gardens would surround the communal rain gardens.

A planted gulley would carry water from the river into the 'village blue' (indicated in figure 18) at times of flood. This landscaped hollow is designed to regularly accommodate floodwater, which would slow the flow downstream, and replenish a mix of dry and wetland habitats, providing both high amenity and biodiversity value.

'Space for Water' would be provided through regrading of the riverbank, creating a low-lying flood shelf with soft banks for a range of habitats; and an extension to the Wandle Trail, providing further opportunities for pedestrians and cyclists to enjoy the river corridor.

The 'village green' would provide a flexible informal recreation area with vegetated banks and areas of hard landscaping. The green would also provide future flood storage potential and space for an array of boreholes for ground source heat pumps. A combination of solar electric panels fitted to all available roofs, and biomass combined heat and power (CHP) would provide a district energy system for the new homes and businesses.

Figure 16 - Courtyard rain gardens would slow the passage of rain overland, storing it until the storm has passed.

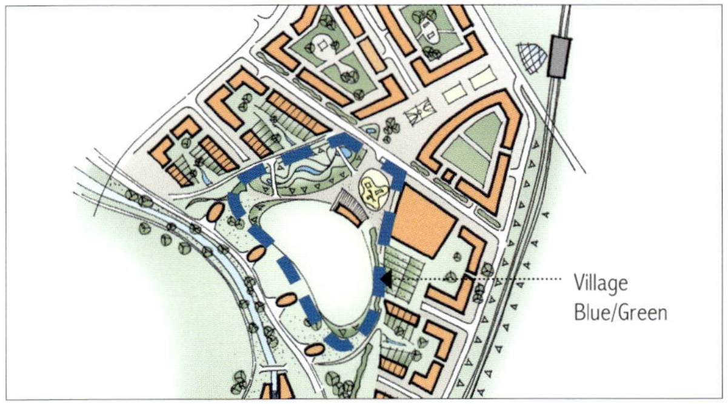

Figure 17 - Site plan

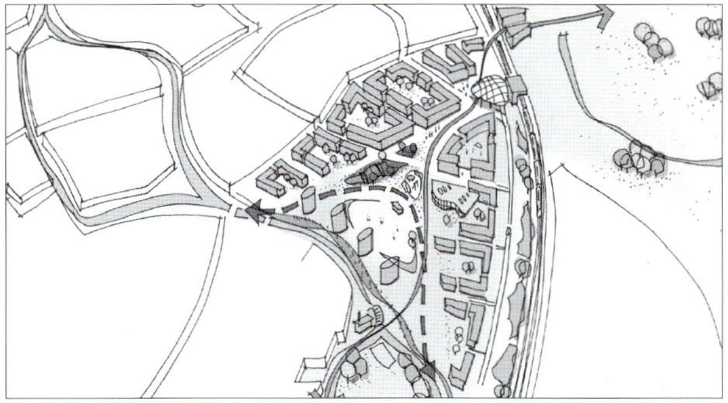

Figure 18 - Aerial view

SITE 2 > The River Nene at Peterborough

Peterborough is located on the middle catchment of the River Nene. Flood risk is predominantly from the river. Floodwater volumes are large and floodwater could remain on site for several days.

Climate change could result in the flood levels and duration of flooding increasing.

Development ideas aim to reunite the city with the river. A mixture of brownfield sites and a landfill site were considered for redevelopment as part of a holistic vision. Higher density development was located close to the city centre and within Flood Zone 1 (low risk). Potential development was organised around a combination of views of Peterborough Cathedral, connections to 'Central Park', and 'stream corridor' drainage paths in times of flood (shown in figure 20).

An adaptable development plan (shown in figure 21) was formed for the Fengate site, a former landfill site. The majority of the site was found to be above the 1 in 100 year flood level, having been elevated by the landfill. Climate change could result in this changing and the area becoming at risk from more frequent events.

Level variations on site would allow drainage and flood paths to be created away from homes. These were envisaged as high quality, wide and attractive green corridors for public gardens and play areas (shown in figure 19). Deeper excavations could provide permanent water bodies creating various wildlife habitats and recreation opportunities. Car parking was located on higher levels of the site and a car club introduced within close walking distance.

The river could be used for delivery of willow coppice for a biomass CHP, located next to a neighbourhood health centre. Solar panels on every roof and over a hundred 15 kW wind turbines located within the stream corridors would be required to meet the electricity demand. A SUDS system, above the flood level, would include green roofs, permeable parking spaces and gravel swales to slow rainwater run off.

Buildings within the 1 in 1000 year flood extent would be designed to be resilient to flooding on the basis that they may be at greater risk in the future. Buildings close to the river would be frame construction with modular panels to allow replacement if required; further away masonry and waterproofing could be used at ground level.

Figure 19 - Development was planned around flood corridors which would guide flood water away from buildings and back into the river once the flood had passed.

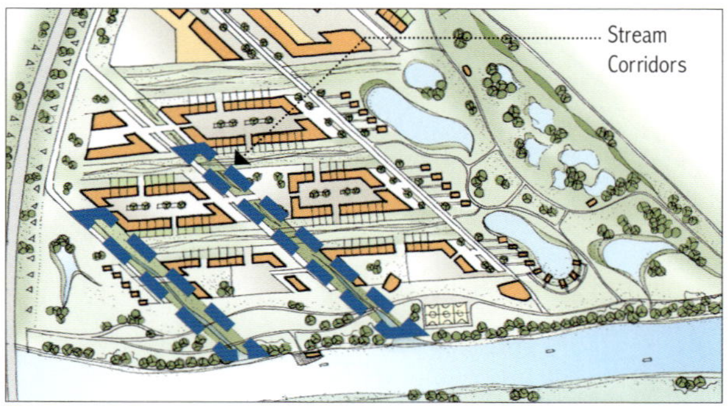

Figure 20 - Site plan

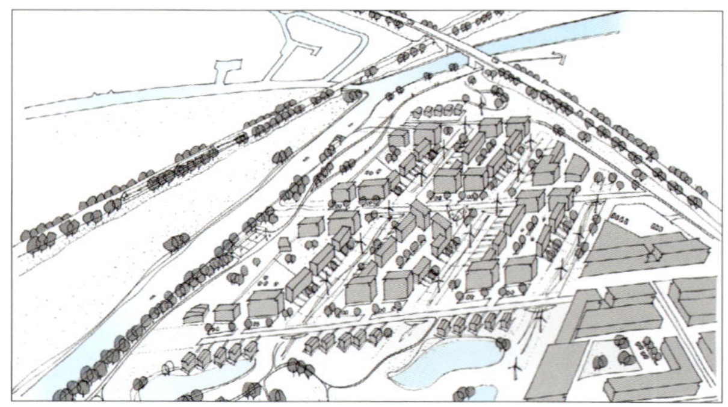

Figure 21 - Aerial view

SITE 3 > The River Arun at Littlehampton

Littlehampton is located on the south coast of England and in the lower catchment of the River Arun. Flood risk is from many sources, the most significant is tidal flooding and overtopping.

The majority of the site is currently Flood Zone 3a, at risk from a 1 in 200 year sea flood. The most significant impact from climate change would be sea level rise and increased storminess. Sea level rise in excess of one metre would change some areas from Flood Zone 1 to Flood Zone 3a. Higher sea level rise would result in much of the south coast falling under high flood risk.

At Littlehampton a concept (shown in figure 23) was developed that would provide the first stage in a long-term, sea-level-rise-linked flood risk management plan for the River Arun and its regions. A regional strategy was suggested to use and enhance the historic floodplain to create a control system, capable of directing storm water away from towns and into dedicated flood storage areas.

The harbour site, currently a 'high probability'[9] flood risk area, could be elevated on spoil from the excavation of the new, 'tidal lagoon', flood storage areas. A mixed-use development could be planned on higher land. Within the lagoons, land could flood during each tide, creating an opportunity for mixed and transitional intertidal habitats, such as deep ponds, tide pools, mud flats, salt marsh, reed beds and grazing marsh. The landscape could provide sufficient distance from settlements to locate large-scale wind turbines (shown in figure 24).

Twelve tidal turbines could be installed within channels designed to direct the flow of water in and out of the lagoons, generating energy for five hours during each successive tide. Wind turbines, solar hot water panels, aquifer heat exchange and tidal turbines could enable all energy demand to be met on site without using solar electric panels or biomass power (an over-subscribed resource). Roofs could then be used for rainwater attenuation.

The Lagoon Promenade could provide multiple opportunities for water-based recreation, amenity and feature floating developments, such as restaurants, boathouses and gardens. Higher density development is concentrated away from the river and elevated above potential flood levels (shown in figure 22). Green roofs and underground rainwater harvesting tanks limit run off, particularly important during high tides when drainage can be inhibited.

Figure 22 - Development is set back from the river behind landscaped river banks. Landscaping provides rainwater attenuation and level variation to reduce the flood impact, should the river defences fail or be breached.

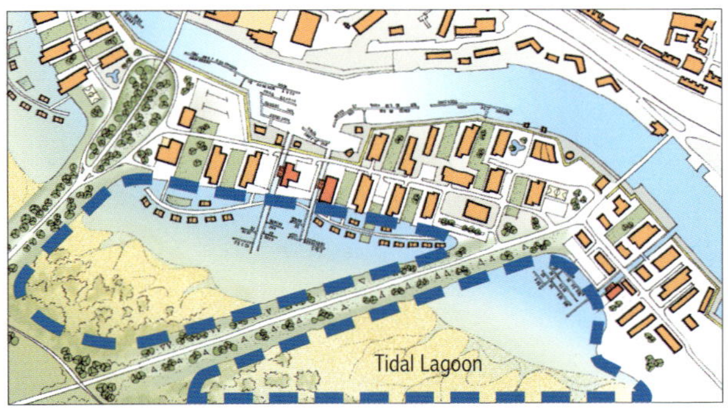

Figure 23 - Site plan

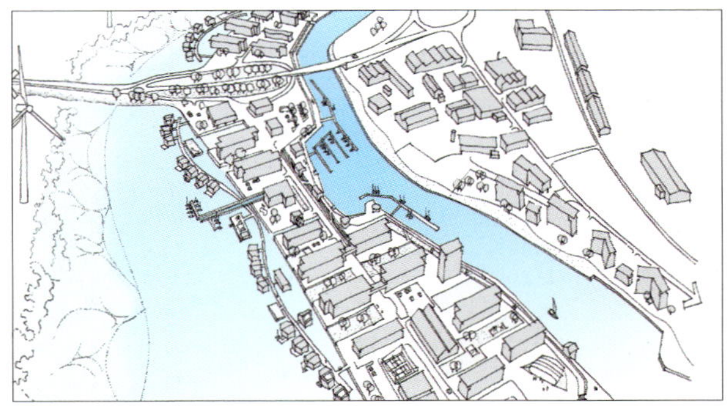

Figure 24 - Aerial view

Summary of Issues

The following section of the report summarises the results of the case studies.

Table 2 describes the sources of flooding, development targets and resulting energy demands for each of the case study sites.

Table 3 lists the means of managing flood risk, the renewable energy technologies used to make the development zero carbon, and identifies the benefits of adopting an integrated design approach.

Location	Site 1. The River Wandle at Hackbridge Upper Catchment Site Area - 13 Ha (Felnex site)	Site 2. The River Nene at Peterborough Middle Catchment Site Area - 50 Ha (developable)	Site 3. The River Arun at Littlehampton Lower Catchment Site Area - 56 Ha (developable)
Flood Risk	The main risk of flooding occurs as a result of intense rainfall. This can overwhelm the infiltration capacity of the land, the discharge capacity of the drains and capacity of the river channel, resulting in flood risk from these sources. Flooding from the land Flooding from rivers Flooding from sewers (theoretical)	The main risk of flooding occurs as a result of an accumulation of flows (caused by rainstorms) from the upper catchment of the river resulting in large volumes of water overwhelming the river channel. The risk of flooding from the land or sewers is possible but not shown in practice. Flooding from the land (theoretical) Flooding from rivers Flooding from sewers (theoretical)	Flood risk to the site is dominated by coastal conditions although fluvial flows could also be significant during a severe event. Extremely large volumes of water are carried by high tides and storm surges. Pluvial flooding can also result when high tides prohibit drainage. Flooding from the land (when tide locked) Flooding from rivers Flooding from ground water (theoretical) Flooding from the sea Flooding from sewers (when tide locked)
Development	1800 homes (900 on the Felnex and rail sites) 65% flats and 35% houses (by area) 7,450 m^2 commercial* (light industrial, office and retail) 2,500 m^2 civic* (health, education and community)	2400 homes (1000 on the Fengate site) 55% flats and 45% houses (by area) 26,832 m^2 commercial (light industrial, office and retail) 4,128 m^2 civic (health, education and community)	2200 homes (1200 on the Harbour site) 53% flats and 47% houses (by area) 23,735 m^2 commercial (light industrial, office and retail) 3,650 m^2 civic (health, education and community)
Energy Demand	 The total energy demand for the Felnex and rail sites was calculated to be approximately 8 GWh/yr, and split: 15% space heating 6% cooling 29% hot water 50% electricity	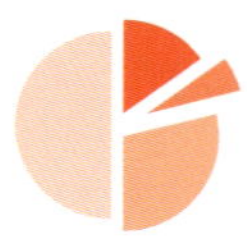 The total energy demand for site 2 was calculated to be approximately 25 GWh/yr, and split: 15% space heating 6% cooling 29% hot water 50% electricity	 The total energy demand for site 3 was calculated to be approximately 22 GWh/yr, and split: 15% space heating 6% cooling 29% hot water 50% electricity

Table 2 - Summary of the case study site issues

Summary of Solutions

	Site 1, The River Wandle at Hackbridge	Site 2, The River Nene at Peterborough	Site 3, The River Arun at Littlehampton
MSW - means of managing flood risk	Flood Risk Management — Avoidance (most vulnerable uses located in lowest risk areas) — Floodplain restoration — Floodplain compensation — River naturalisation Sustainable Drainage — Gravel and planted swales and rain garden detention basins	Flood Risk Management — Avoidance (most vulnerable uses located in lowest risk areas) — Floodplain conveyance paths — Floodplain compensation — Elevated and amphibious construction — Land set aside for future floodplain Sustainable Drainage — Green roofs, gravel swales, permeable paving	Flood Risk Management — Avoidance (most vulnerable uses located in lowest risk areas) — Floodplain restoration — Managed realignment — Controlled inundation areas (regional and site level) — Raised river banks — Elevated and resilient construction — Safe haven buildings — Improvements to the flood warning system Sustainable Drainage — Green roofs, gravel swales, permeable paving
Zero carbon technologies	— 49% Biomass CHP and boilers — 25% Solar photovoltaic panels (PV) — 13% Energy service company (off-site)* — 7% Borehole ground source heat pumps (GSHP) — 6% Solar thermal panels * Plus a small contribution from micro hydro	— 45% Biomass CHP and boilers* — 27% Solar photovoltaic panels (PV) — 12% Wind turbines — 11% Aquifer thermal energy storage (ATES) — 6% Solar thermal panels * Plus a small contribution from biogas	— 66% Wind turbines — 11% Solar thermal panels — 9% Aquifer thermal energy storage (ATES) — 8% Borehole ground source heat pumps (GSHP) — 5% Tidal turbines
Wider benefits	— Habitat creation — Village centre and sense of identity — Access to the river — Downstream improvements — Riverside frontages	— Habitat creation — Improved access to river — Improved pedestrian connections — Riverside frontages	— Habitat creation — Regional reduction in flood risk particularly to Littlehampton — Access to the river — Public open space and recreational facilities — Local economic opportunities — Unique place making
Benefits of an integrated design approach	— Shared land use allowed planning requirements to be met simultaneously, resulting in higher numbers of homes. — Housing courtyards provided play areas, gardens, rainwater attenuation and harvesting. — Use of floodplain for recreation helped to approach amenity standards therefore allowing more homes on the site. — Location of GSHPs within the floodplain enabled use of this technology, otherwise difficult on high-density developments. — Integration of swales into the whole site created visual stimulus and flood risk awareness.	— Flood corridors provided sufficient space to locate and partially conceal wind turbines on the development site. — Corridors provided play areas, gardens, allotments, floodplain conveyance paths and movement corridors. — Use of floodplain for recreation helped to approach amenity standards therefore allowing more homes on the site. — Location of wind turbines within the corridors enabled suitable distances from the buildings to be provided. — Integration of planted corridors into the whole site created visual stimulus and flood risk awareness.	— Extensive floodplain restoration provided space to locate large efficient wind turbines away from development. This provided scope for renewable energy to wider area as well as the zero carbon development. — Tidal lagoons, set aside for flood risk management, could also be used for energy generation and habitat creation. — Reduction in PV demand allowed green roofs to be considered in floodplain areas. — Flood risk measures allowed regeneration of the harbour site and economic contribution to the town.

Table 3 - Summary table of the solutions proposed for the case study sites

Results and Findings >

Managing Flood Risk

In recent years the emphasis of coping with flooding has shifted from defence to risk management. An over-riding objective in setting present day flood and coastal erosion risk management (FCERM) measures is to work with natural processes and avoid the construction of 'unnecessary' flood defences, particularly where these defences would remove areas of fluvial or coastal floodplains.

The risk of flooding was identified from current and historic maps, site inspection and local advice. Information on flooding from the river and seas was obtained from the Environment Agency and included the emerging information from the CFMPs and SFRAs where available. The Environment Agency provided little or no evidence of flooding from the land, sewers or groundwater for any of the sites. However, as a precautionary approach it was assumed that flooding could occur from some of these sources.

Development and flood risk

It was found that flood risk could be managed without the need for flood defences in each of the sites except for parts of site 3, Littlehampton. On each site, areas of land existed in all four flood zones therefore a sequential approach to site selection and development use could be applied.

Figures 25 and 26 illustrate the respective current and future flood zone maps for site 1, Hackbridge. Figure 24 indicates that the area nearest the station is Flood Zone 1, the majority of the site Flood Zone 2 and only a small area Flood Zone 3a. Therefore residential (more vulnerable) use of the site would be considered appropriate within PPS25. The possible future extent would suggest that less vulnerable uses would be more appropriate.

The majority of the development proposed on site 1 could be located above the 1 in 100 year flood level, (outside Flood Zone 3). The remaining property could be designed to resist the relatively low depths of flood water that may occur and provide safe access.

The small upstream catchment and 'flashy' peak river flows (as indicated by the flood hydrograph, Appendix A) suggested that flood storage to attenuate peak flows could be an option. However, the lack of available sites (due to the current extent of development within the river catchment) precluded any obvious opportunity. Therefore the village blue/green was conceived as a small flood store, into which floodwater would be channelled through the site in a controlled manner to slow the passage downstream. Regrading of the river banks to create soft planted edges would provide space for the river to flood naturally, reducing maintenance requirements and providing environmental enhancement.

On site 2, Peterborough, the majority of the site was found to be above the 1 in 100 year flood level and only parts of the site, mostly within the city park, were found to be lower than the 1 in 1000 year flood level. The change in flood flows that could occur as a result of climate change showed that much of the site could be at risk of flooding in the future, particularly from the 1 in 1000 year event. Development could therefore be located above predicted river flood levels but less easily for future events.

Flood depths were such that as a precautionary approach small changes to the land levels could be implemented to raise properties above the future 1000 year event and provide compensation with some areas of land on site. In areas closer to the river a precautionary approach would be to make buildings resilient.

On site 3 the depth, speed and volume of water that would result from tidal flooding were such that development of the majority of the total land area was not suitable. However, the scale of the site meant that there were areas above the 1 in 200 year and 1 in 1000 year flood level that could be prioritised. The harbour site was predominantly within flood zone 3, therefore redevelopment of the site would need to be appropriate and safe. The availability of land meant that strategic flood management measures to reduce flood levels were possible. However, flood defences or land raising would still be required, especially considering the possibility of increased flood levels resulting from sea level rise. Therefore flood risk management measures included regional flood storage, land raising, defences and resilient construction. The use of multiple management measures was found to provide the most cost effective, adaptable, robust and attractive solution.

Figure 25 - Site 1 current flood risk map (based on draft modelling information 2008)

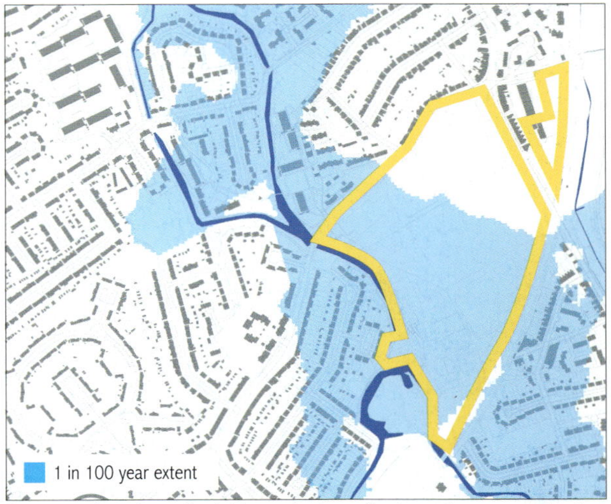

Figure 26 - Site 1 future flood risk map (no future flood zone 2 extent was given)

Managing Flood Risk

Sustainable drainage

Sustainable drainage systems (SUDS)[36] are not designed to manage flood risk from the river or the sea. SUDS are an alternative to conventional surface water drainage systems that cope with rainwater run off. SUDS can help to reduce the likelihood of drains and sewers overflowing by reducing the amount of water entering the systems and can be designed to cope with all rainwater run off, avoiding the need for conventional surface water drains.

Conventional sewers are typically designed to deal with relatively common rainfall events (such as less than 1 in 30 year rain storms). Therefore these are likely to be overwhelmed during a 1 in 100 year storm event. The land area required for a shallow SUDS bio-retention system to attenuate rainwater from a 1 in 100 year rainstorm event was found to be between 5% and 10% of the total site area for each of the study sites.

In areas at risk of river flooding from a 1 in 100 year storm event attenuation was provided in the form of green roofs to avoid rainwater falling on flooded ground. Within the rest of each site attenuation was provided at ground level in the form of swales, detention basins and rain gardens.

The ground level rainwater attenuation for each site could be integrated within the development, providing space for recreation and environmental gain.

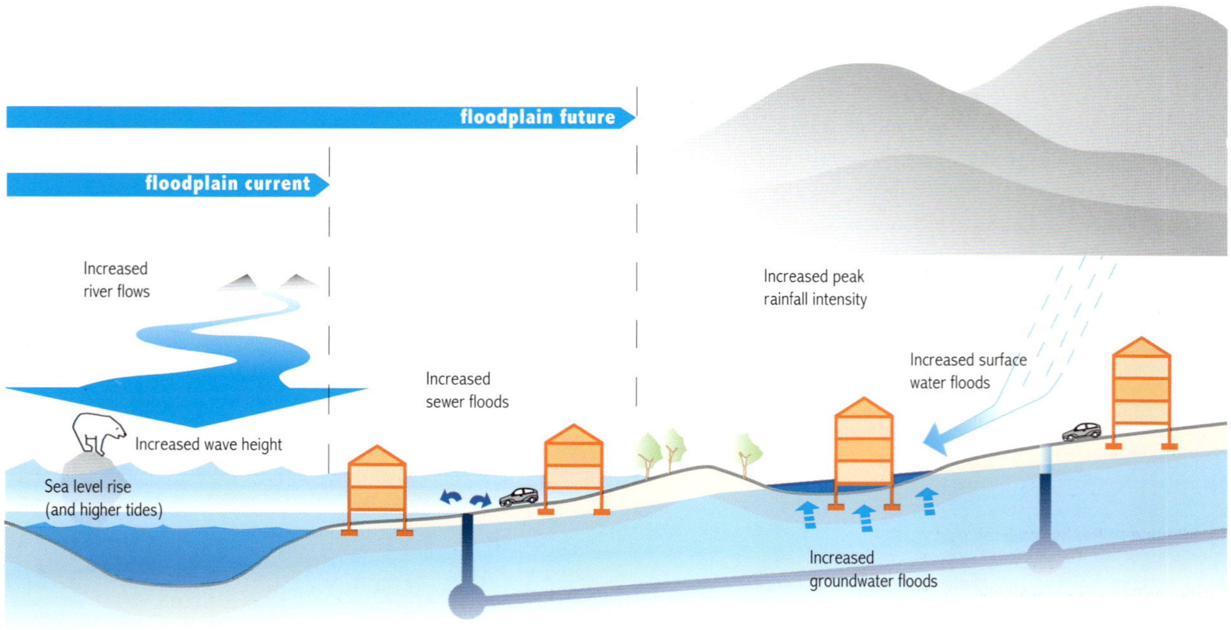

Figure 27 - Potential effects of climate change

Adapting to climate change

Using the guidance set out within PPS25 annex B it is possible to establish the changes to the extent of flood risk that could result from climate change. The precautionary allowances set out in tables B.1 and B.2[9] were used to assess the future events (for the period 2085 to 2115) for each site. Figure 27 illustrates the implications of climate change on the hydrology of the UK (increased rainfall, storminess, wave height and sea level) and would effect the study sites.

In each of the three study sites it was considered that the predominant impacts of climate change on each site could be identified as:

Site 1 — Increased peak rainfall intensity would increase rates of run off, increasing the risk of flooding from the land and sewers and result in more flashy river flooding.

Site 2 — Increased river flows would increase flood levels and duration. Increased peak rainfall intensity would increase the risk of flooding from land and sewers.

Site 3 — Increased sea level and increased storminess would result in higher storm levels leading to a risk of banks and defences being overtopped. Increased river flows would also increase river levels. Increased peak rainfall intensity would increase the risk of flooding from land and sewers.

A precautionary approach was applied to all three sites and designs were proposed that could be adapted to future extreme events.

Energy

Meeting Zero Carbon

To achieve a zero carbon new build development (a requirement of Code Level 6[3]) all of the energy demands of the development should be provided by renewable energy produced on site.

The scale of the developments (over 1000 homes) considered for the case studies enabled a range of renewable energy technologies to be considered. However, it was found difficult to achieve sufficient energy from on-site resources on the case study sites with the exception of site 3, where a greater abundance of energy sources were available.

There were several reasons why the target proved difficult to achieve, however, the main reason was lack of space.

Though technically feasible to use biomass to provide 100% of all energy demands on site, sole use of biomass combined heat and power (CHP) would waste fuel. Excess heat would be produced when fuel is burned to meet electricity demand, particularly during the summer. If this heat could not be used it would be 'dumped' into the environment, contributing to the 'urban heat island' effect and potentially increasing the need for mechanical cooling, further increasing energy demands. Carbon miles from the transport of biofuel provides a further reason for careful consideration of biomass energy in urban areas. Therefore, a range of energy sources was explored on each site to identify the most efficient integrated energy solution.

Energy Split

Energy demand for a development is subdivided into four components: heating, cooling, hot water and electricity (light and power).

Historically, space heating accounted for almost 50% of energy demand within the home, as shown in figure 28a. Increased thermal performance standards set by building regulations have reduced the energy required for space heating, and subsequently the overall energy demand, as shown in figure 28b.

The Code Level 6 standards for thermal resistance and air-tightness exceed building regulations (Part L 2006). This reduces the energy required for space heating, as shown in figure 28c.

To further reduce overall energy demand the remaining needs must be targeted. Though the demand for electricity will be reduced by energy saving appliances it is the main demand of new homes.

Due to the similarity in the density and accommodation mix of each of the proposed concept plans the split in energy demand for each site was also similar. Electricity demand was found to make up approximately 50% of the total demand for each site.

Energy Demand

To meet all energy needs different energy sources were required. For example, wind turbines generate electricity but do not provide hot water. Biomass CHP is often favoured over other energy sources due to lower capital costs, lower space requirement and versatility in energy production. However, a community energy system such as CHP is generally only viable for medium to high-density masterplans of 500 dwellings or more.

For both sites 1 and 2 biomass was selected to provide nearly 50% of the total energy demand. To meet the additional electricity demand the most suitable technologies are wind and solar PV.

On site 1 wind turbines were found unsuitable, therefore solar PVs were required. However, the area required exceeded the available roof area. Therefore to achieve a zero carbon development the remaining electricity would need to be purchased from off-site renewable sources.

On site 2 wind turbines were technically feasible but local planning constraints were likely to preclude large turbines. Therefore a combination of smaller turbines and solar PVs was explored. It was found that all available roof area would need to be used for solar PVs to provide the remaining electricity demand.

On site 3 large wind turbines were technically feasible, therefore it was proposed to use wind turbines to generate 44% of the electricity, with biomass CHP and tidal turbines used to generate the remainder. Alternative options explored were: biomass CHP with solar PVs and wind, and a wind option that used electric heating and required no biomass.

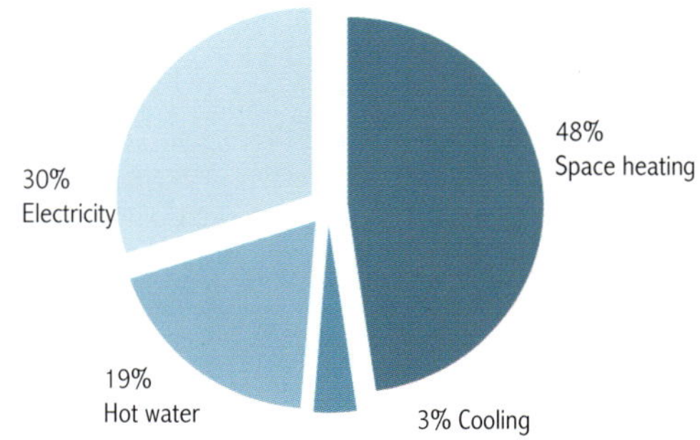

Figure 28a - Energy split for traditional development based on 85% domestic, 15% commercial (based on existing stock)

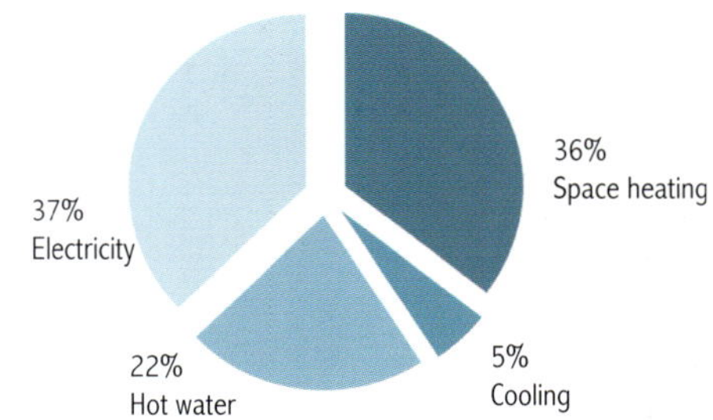

Figure 28b - Energy split for development to Part L 2006 based on 85% domestic, 15% commercial (total demand is 33% less than existing stock)

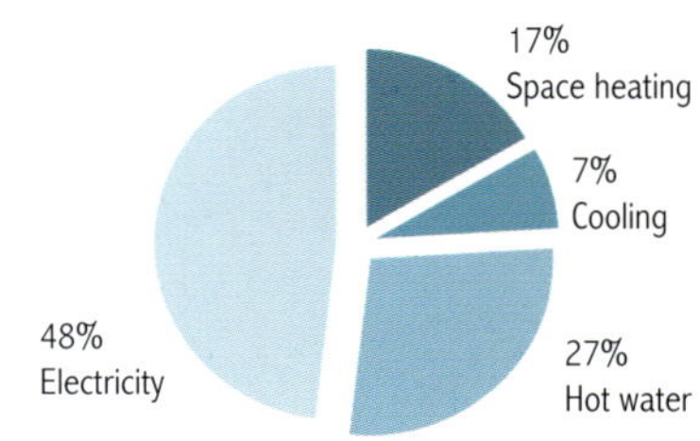

Figure 28c - Energy split for zero carbon development based on 85% domestic, 15% commercial (total demand is 52% less than existing stock)

Energy

Space for Energy

Solar panels (PV and thermal), ground source heat pumps, wind and tidal power all require considerable amounts of space to locate and function.

The area of solar panels required on site 1 was found to be greater than the total viable roof area of the development and on site 2 a large number of wind turbines were required to meet the remaining electricity demand. Solar panels fitted on other surfaces were likely to be less efficient. Therefore higher density development would have had insufficient space for solar panels.

Based on a 3 m grid of 110 boreholes the ground source heat pump system at site 1, Hackbridge, would cover 8100 m^2.

Large wind turbines can create vibration, noise and shadow flicker. Siting requires considered assessment but typically turbines would need to be located upto 500 m away from homes, roads, receptors and other turbines. Therefore they are typically only feasible on large sites or sites near to areas of open space, as in site 3, Littlehampton.

Tidal power requires large volumes of water to generate significant power outputs. This requires large areas of land to be inundated with water or areas of rivers to be cut off from the main channel. On site 3, the flood storage area used for energy generation was 50 hectares, equivilant to the entire developable site area.

Therefore on higher density development sites these energy sources are able to produce an increasingly smaller proportion of the site's energy needs. Biomass is currently considered as an on-site solution even though only the technology is on-site, not the fuel. When the area required for the biomass crop is included it becomes far less viable on-site and highlights the need to encourage the uptake of other energy sources to avoid increasing pressure to use land currently used for food crops.

Figure 29 compares the site area and space required for the various energy technologies proposed to achieve a zero carbon development at Hackbridge. If the buildings were built to current building regulation standards the energy demand and thus the area required to achieve zero carbon would increase. The average building height is approximately 4.2 storeys. The area of land required to grow the amount of biomass required (based on a short crop rotation and three harvests a year of willow) is five times the total area of the site and 24 times the area of the buildings.

Summary of Energy Issues

— Very high thermal performance buildings are required to reduce overall energy demand.

— A range of energy technologies are required to meet energy demand. These need space, particularly solar panels for electricity and hot water generation.

— Higher density sites are unlikely to be able to provide this space therefore they are likely to require off-site energy generation but can offer other sustainability benefits through improved access and transport.

— Building surfaces should be designed to maximise solar orientation for solar electricity generation (PVs) and solar water heating (thermal). At the same time buildings should be designed to promote natural cooling.

— The number of available energy sources such as wind and tidal may increase further down river. Associated energy technologies can be used to reduce reliance on biomass for renewable energy and can be integrated into the spatial plan and other uses.

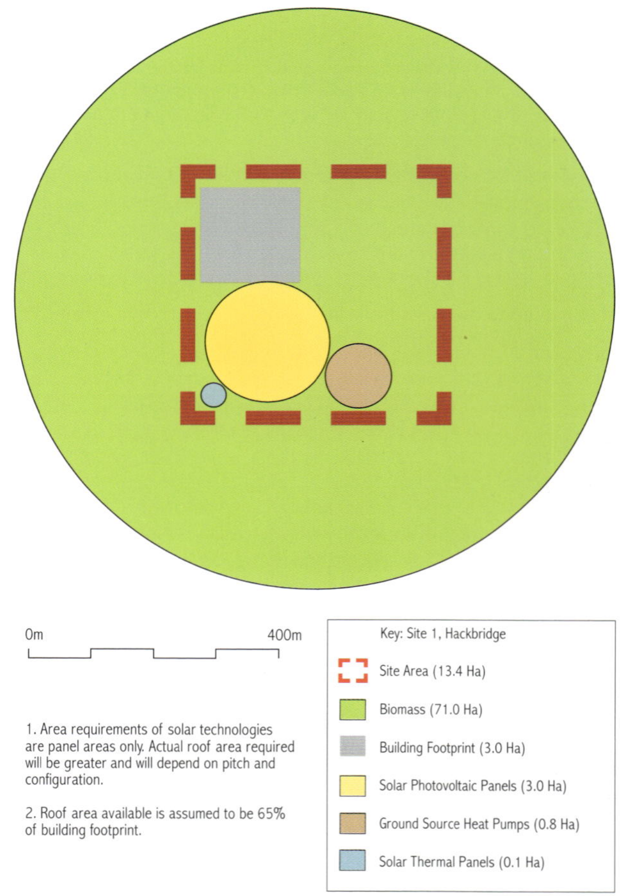

1. Area requirements of solar technologies are panel areas only. Actual roof area required will be greater and will depend on pitch and configuration.

2. Roof area available is assumed to be 65% of building footprint.

Figure 29 - Space requirements for renewable energy technologies required to achieve zero carbon development at site 1, Hackbridge

Development and Amenity

Providing Space for Amenity

The Six Acre Standard[5] was established in 1925 (and revised to include children's playing space standards in 1992) to determine minimum standards for outdoor playing space. The LifE proposals attempted to meet the six acre standard and any additional local planning requirements through on-site amenity provision.

Figure 30 illustrates the outdoor space required for site 1 residents based on local guidance (totalling approximately 64 m^2 per person). This illustrates that the total amenity required was greater than the total site area. The lack of adequate local outdoor play areas, exacerbated the need to provide as much amenity on-site as possible. The 'Village Blue/Green', courtyard rain gardens, and naturalised river edge helped to provide 46,000 m^2 of public open space, parkland and sports provision (approximately a third of the total 141,000 m^2 required in local standards). By designing these areas to cope with flood water, such as resilient playgrounds, much of this area could be maintained as floodplain or used for water attenuation.

On sites 2 and 3 where locally set PPG17[32] compliant standards did not exist (at the time of writing) it was difficult to reach confident predictions on the requirements. However, based on the six acre standard approximately 36 m^2 of open space was required per person. This equated to approximately 50% of the developable area for each site.

In each of the sites, use of the floodplain as part of the open space was the only feasible solution to meeting the open play standards. This included the Village blue/green on site 1, Peterborough central park and stream corridors on site 2 and the proposed recreational wetland park on site 3.

PPG17 states that 'local authorities should...recognise that most areas of open space can perform multiple functions'[32]. It does not specifically identify flooding or energy as an aspect of multi-functionality, however PPS25 identifies the need for multi-functional use of the floodplain.

Whilst 'key local provision should not be located on land liable to flooding', it is possible that other local provision can, particularly on land with a low probability of flooding. A design-led approach, as pursued in the LifE project, could promote suitable solutions through negotiation with local planners.

Development Density

Each concept plan was based predominantly on providing residential development with an appropriate level of commercial and civic development.

On the basis of the respective local planning guidance for each location the densities of the development ranged from approximately 40 units per hectare to 75 units per hectare. This is generally considered medium density development.

Higher density developments are sometimes considered essential to sustainable design on the basis of reducing car use, increasing energy efficiency and preserving land for crops. However, greater numbers of homes and people require more energy and outdoor space to be provided on the same area of land.

The conceptual masterplans demonstrate the difficulty in providing sufficient space for energy, outdoor play and for SUDS without multi-functional land use. When the additional space requirement for floodplain was factored in multi-functionality became essential.

On higher density developments multi-functional land use is likely to be essential, regardless of proximity to the floodplain.

The space required to allow land to flood offered a good opportunity to provide additional use for amenity and renewable energy infrastructure.

Providing Space for Wildlife

Many derelict brownfield sites contain an abundance of wildlife and habitat. New development can threaten and cause harm to this habitat, especially during the construction work. Creating space for wildlife to cohabit with new occupants adds further demand on the available land. Wildlife is very compatible with wetland areas and riparian borders associated with natural floodplains; therefore flood mitigation that works with natural processes can help to improve habitat provision.

Figure 30 - Space requirements for outdoor amenity at Hackbridge

Costs

LifE Costs over Current Standards

The cost of providing space for water and the cost of building to Code for Sustainable Homes Level 6, in particular, were found to increase the capital costs over traditional development. However, an integrated approach was found to reduce these costs and provide more extensive environmental benefits than a non-integrated design.

Figure 31 (opposite) shows the cost of developing each site to Code Level 3 (referred to as the base case) compared with the cost to develop each site to the LifE standard (Code Level 6 with non-defensive flood risk management).

- Site 1, 34% increase in total development cost over base
- Site 2, 32% increase in total development cost over the base*
- Site 3, 11% increase in total development cost over the base**

The majority of this cost increase was a result of the increased energy and utilities costs required to build to Code for Sustainable Homes Level 6 and BREEAM Excellent.

The cost of providing space for water was found to be cheaper than providing flood defences and compensation. It represented 1% of the total development cost of sites 1 and 2 and 9% of the total cost of site 3. The cost of improving certain buildings to make them flood resilient was found to increase the total development cost by between 5% and 7% depending on the location, therefore negating any substantial saving. However, the cost of installing, even expensive, flood resilience measures can be recouped with a single flood.[37]

The Association of British Insurers (ABI) reported that 'flood claims typically reach £15,000 to £30,000'[38] and averaged £20,000 in the 2007 floods, with repairs taking between 12 and 18 months.[39] The economic cost of this period of disruption can be far greater than the cost of the repair work. The ABI are therefore calling for standards of resilience to be incorporated within the Building Regulations.

'The UK market is unique in providing All Risks insurance including fire, theft, storm, flood, escape of water, subsidence, property owners' liability and, in most cases, accidental damage as standard cover'.[38] Future changes may see premiums increased for properties in high flood risk areas and subsequent reductions in those premiums for properties incorporating resilience measures.

The cost of flood risk management was found to be more expensive the lower down the river catchment the site was located. However, the range of renewable energy sources increased, particularly on site 3, where the proximity to the coast increased the windiness Overall this meant that the costs on site 3 (lower catchment) could be reduced by the selection of more economic energy technologies, and the feasibility of making the site zero carbon improved.

It is expected that the cost of renewable energy technologies will reduce in the future through increased production to meet demand and technological advances. This should help technologies such as PV, GSHP and tidal to become more cost effective, improving the options to meet the energy demand on site.

Reduced car dependency through the provision of car clubs and pedestrian access to public transport connections was shown to reduce the amount of roads required, particularly on site 1, which was closest to the main line rail station. The remaining space was then suitable for landscaping and planting, introducing greenery within the development, adding attraction and potentially reducing the urban heat island effect. This approach was found to reduce the total development costs by between 1% and 5% on the study sites.

---

* Site 2, Option 2

** Site 3, Option 2

Costs

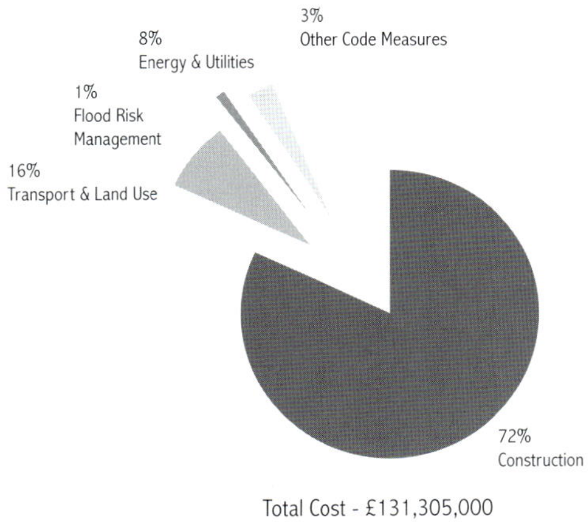

Site 1 - Cost of base case

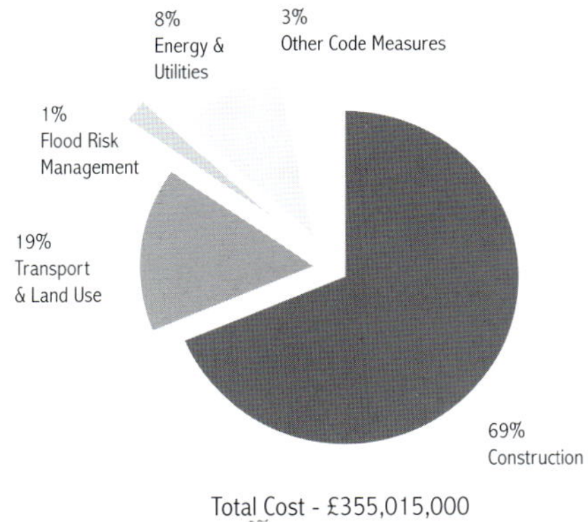

Site 2 - Cost of base case

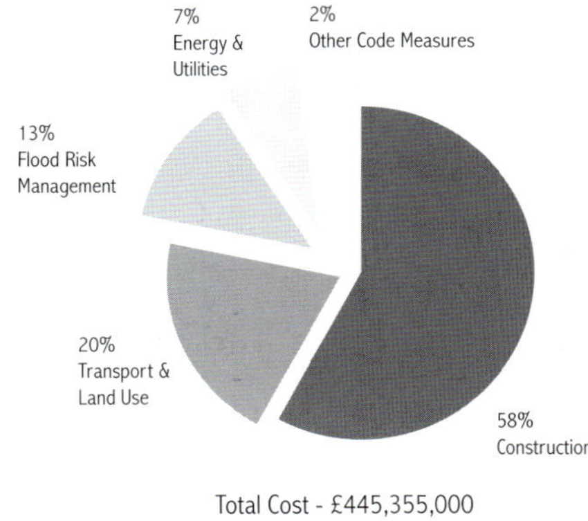

Site 3 - Cost of base case

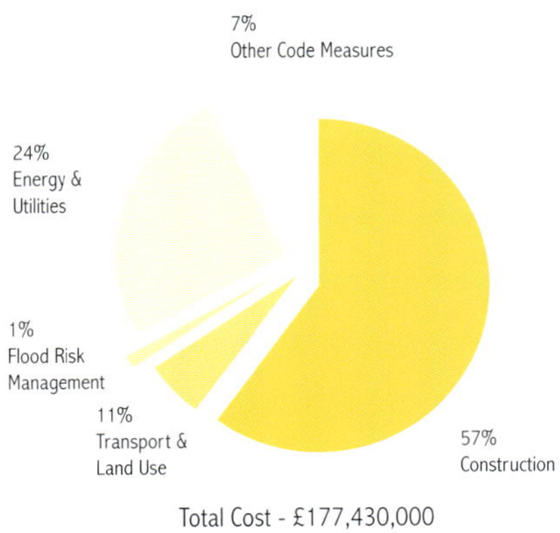

Site 1 - Cost of LifE proposal

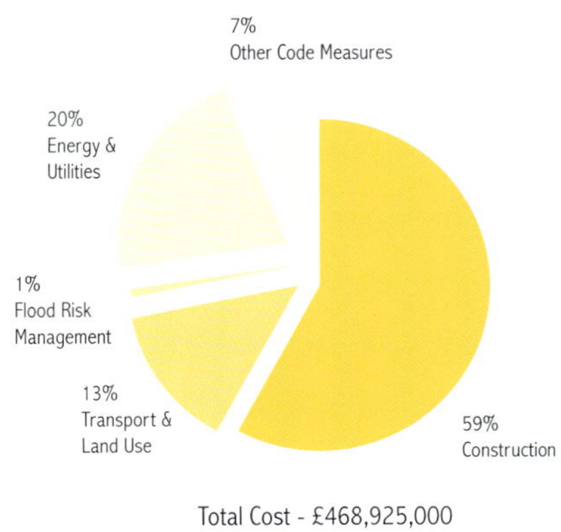

Site 2 - Cost of LifE proposal (option 2)

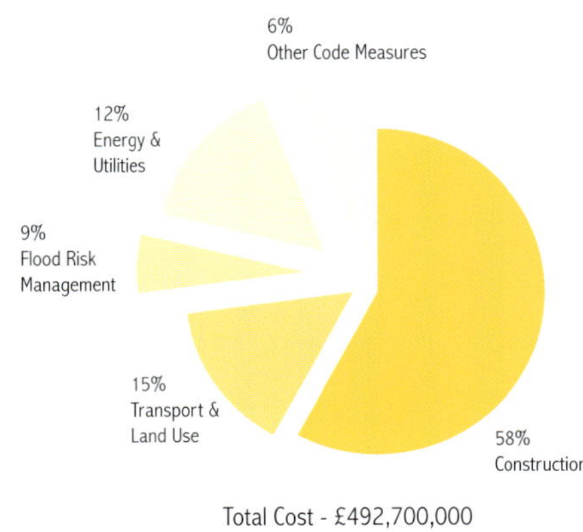

Site 3 - Cost of LifE proposal (option 2)

Figure 31 - Cost comparison between base case and LifE option for the case studies.

Costs

Integrated Approach

To achieve Code Level 6 and provide space for water separately, the capital cost compared to the base cost was found to increase by 35% on site 1, 37% on site 2, and 21% on site 3 (see figure 32).

The LifE approach resulted in an increase in capital cost over the base cost of 34% on site 1, 32% on site 2 , and 11% on site 3. This represented a saving in achieving these standards by adopting an integrated approach.

It was assumed that certain components of the development would not be viable without adopting an integrated approach and other must be installed separately, therefore increasing the overall cost. For example, wind turbines could not be accommodated without the large clearings proposed in the LifE options, therefore greater numbers of PVs were required. Conventional car parking and road provision were assumed to be required to suit conventional planning. Deep swales were assumed to be needed for sustainable drainage, rather than shallow accessible rain gardens, rain corridors and rain courts (used in the LifE proposals for sites 1, 2 and 3 respectively) which would require a greater land area. A more detailed assessment may also identify the requirement for green roofs, which would further increase the cost.

The 'LifE approach' was found to provide greater opportunities for more cost effective technologies and savings through efficient planning. The most significant savings resulted from the use of wind turbines in preference to solar PVs. On site 2 wide spacing between blocks of housing was required to provide space for medium scale wind turbines (this complemented the aim to reduce development footprints within the future floodplain). On site 3, large-scale wind turbines could be located within the flood storage areas, well away from development. Localised regions of higher density development also helped to reduce walking distances and car parking requirements on all three sites.

Land Receipts

Land value is commonly calculated by assessing the capital cost of a development, the required yield from that development and the projected market value of those properties built upon that development. The results of this assessment will provide a land value which the developer is willing to pay to obtain that particular piece of land, the land receipt.

The riverside locations may help to increase market values by 5% to 15%. However, this is unlikely to cover the increased capital cost. Therefore the increased capital cost is likely to reduce the developer's return or the land receipt offered to the land owner. This short-term view that land values will be dramatically reduced may create an unwillingness amongst land owners to offer land for development of this type. This could then ultimately lead developers to continue using a traditional approach as development costs and projected property values create certainty of return.

The number of planning applications that have been approved against Environment Agency advice has been decreasing in parallel with their increased consultation role. Future planning applications are likely to require the approval of the agency, this may reduce the likelihood of traditional proposals obtaining planning permission in flood risk areas. In turn this may reduce the value of the land.

As higher standards of construction and sustainability are made obligatory for new developments the value of land has the potential to decrease. This decrease in value could result in land becoming scarce as owners will be reluctant to accept lower receipts.

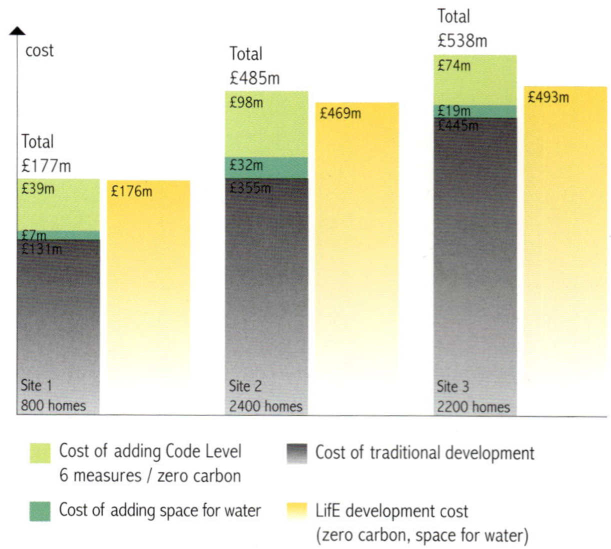

Figure 32 - Comparison cost of un-integrated and integrated design approach

Costs

Lifetime Costs

When costs were considered over the lifetime of the development (assumed to be 100 years) the costs of a traditional development were projected to be greater than that of a LifE development. The cost of future energy supply and material enhancements that may be required for a traditional development become less certain over time.

Renewable energy sources reduce dependence on imported fuel and may help to reduce cost uncertainty. Adaptable and resilient development design may also mitigate expensive retrofitting costs. The more intense the effects of climate change the more quickly the payback will be realised. Figure 33 illustrates how the uncertainty of payback will be realised. Figure 33 illustrates how the uncertainty of future energy prices and increased flood risk could lead to uncertainty in accurately determining the payback period for capital investments made at the time of construction.

Energy Service Companies

The involvement of Energy Service Companies (ESCOs) has the potential to enable developers to offset some of the initial costs of renewable technologies, with the service provider making their return through a long-term supply contract. Consultation with ESCOs previously indicated that the minimum scale of development which would sustain a business case for ESCO involvement would be in the order of 300 dwellings. This number may be slightly reduced on schemes with a non-residential element such as the sites used within this report.

A high level business case exists for ESCO involvement, which could offer long-term fuel bill reductions to occupiers, and capital contribution toward the initial installation of the plant and replacement at the end of its useful life. This is achieved by considering the marginal discounted maintenance and replacement costs together with marginal revenues over a minimum of 35 years.

Implementation

From previous experience it has been found that developments with uniquely high standards of construction and sustainable solutions on a scale, such as those proposed under the LifE options, have struggled to move forward due to the capital cost of the project and the effect this has had on the availability of land.

The Homes and Community Agency and other (typically public) landowners are often prepared to offer land at a reduced price to enable projects such as LifE to progress, promoting unique and sustainable construction, and high design quality. The provision of land at a reduced price provides an excellent mechanism by which to offset the increase in capital cost.

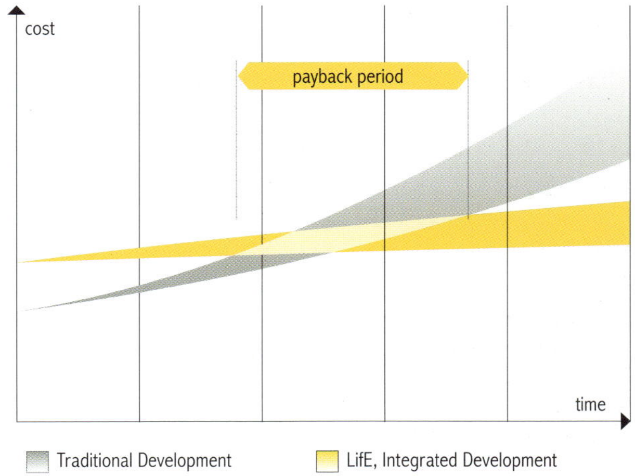

Figure 33 - Potential payback period for LifE versus traditional approach

Transferable Ideas >

Make Space for Water, Energy and Play

Transferable ideas

A number of generic principles emerged from the project work. These are indicated opposite. An over riding need for space to support development emerged, as did the need to use land for multiple functions. Therefore the concept of making space for water can be extended when considering development in flood risk areas to include energy, recreation and amenity use.

Generic Solutions

The case studies illustrated the complexity of understanding flood hydrology in any given context and the difficulty with developing generic concepts. However, some generalities can be inferred on the basis of catchment location and scale, see figure 34.

Within the upper catchment of a river the intensity of the rainfall is likely to dominate the flood risk. Therefore slowing the flow of rain through use of more landscaped surfaces and sustainable drainage should help reduce flood risk within sites and reduce peak flows downstream.

Within the middle catchment of a river the fluvial flood risk is likely to dominate. The importance of slowing rainfall applies, particularly within the upper catchment. Flood water storage requirements are likely to be too great to consider within urbanised areas. Allowing water to pass through the urbanised areas, and reducing obstructions to river flows are likely to be advisable. Designated flow routes will also benefit site recovery after a flood.

Within the lower catchment of a river/estuary the tide is likely to dominate the flood risk. The importance of slowing rainfall applies as does the requirement to avoid hindering river flows. However, the tidal flood risk is likely to dominate. Floodwater storage requirements are likely to be too great to consider within urbanised areas and flood risk is likely to be too great to encourage water onto sites, particularly in low lying areas. Therefore letting tides flow around developments into designated areas and, if defences are breached, creating designated flow routes are likely to be of benefit.

The following pages, from the LifE Handbook, illustrate some practical solutions to possible development in different parts of a river catchment.

Upper Catchment - Let Rain Slow

Middle Catchment - Let Rivers Flow

Lower Catchment - Let Tides Go

Figure 34

LifE Project Principles

Provide space for the river

Create areas that flood, to compensate for any loss of floodplain and to direct floodwater away from homes. Protect the natural floodplain, where possible.

Provide space for rain

Create areas that slow and store rainwater out of the floodplain to reduce surface water run off rates to at least equivalent to greenfield rates.

Create space for amenity

Locate play and recreation areas in areas designed to flood. Use permeable surfaces to reduce run off and help prevent water-logging.

Integrate with community needs

Plan proposals that could complement the area through provision of amenities and services that are lacking.

Design to be adaptable

Identify space to accommodate more frequent and higher volume floods in the future. Allow for retrofitting of resilience measures and new renewable technologies.

Reduce car dependency

Design layouts to encourage walking and cycling, establish car clubs on-site to reduce car ownership to 1 car per 20 people.

Provide space for energy

Orientate roofs to the south for solar panels and make space within development sites for wind turbines where possible.

Provide backups

Design fail safe mechanisms, such as 'safe haven buildings' that are built to a higher resilience standard to provide sanctuary.

Upper Catchment > Let Rain Slow

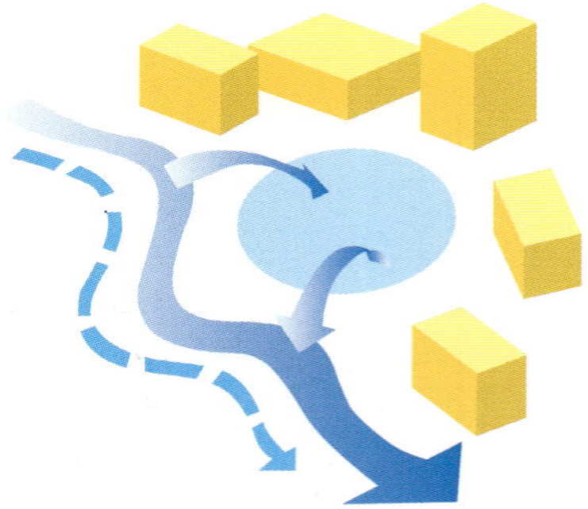

Within the upper catchment of a river the land is typically steep and the catchment area small. When rain falls on the land it runs into the river shortly afterwards causing river levels to rise and fall quickly and with little warning time.

Urbanisation and hard surfaces can exacerbate this problem. Therefore using permeable surfaces and dedicated areas that slow the path of the rain into the river can reduce flooding.

With very small rivers it may be possible to store some floodwater on site and reduce risk further downstream. Re-naturalising river banks can also improve space for the river to expand and create opportunities for habitat.

Community energy centres (such as CHP) located away from flood risk.

Roofs to face south for use of solar panels.

Rain gardens and swales slow rainwater run off. These should be located above the floodplain.

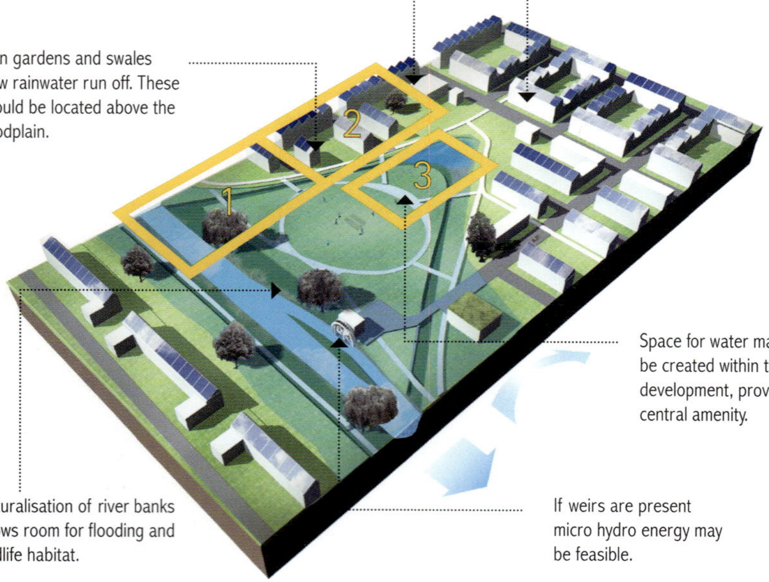

Space for water may be created within the development, providing central amenity.

Naturalisation of river banks allows room for flooding and wildlife habitat.

If weirs are present micro hydro energy may be feasible.

Figure 35

SECTION 1 - Riverside

Private amenity elevated above ground can be used to attenuate rainwater run off, provide views and preserve floodplain.

Communal garden strip provides a buffer to private developments.

Naturalised river banks increase flood storage, create visual amenity and habitat.

SECTION 2 - Rain courtyards

Roofs used for solar PVs and solar hot water.

Planted communal 'rain gardens' provide buffer to local areas of play.

Green swales provide threshold to buildings.

Rainwater is collected within communal harvesting system.

SECTION 3 - Flood storage

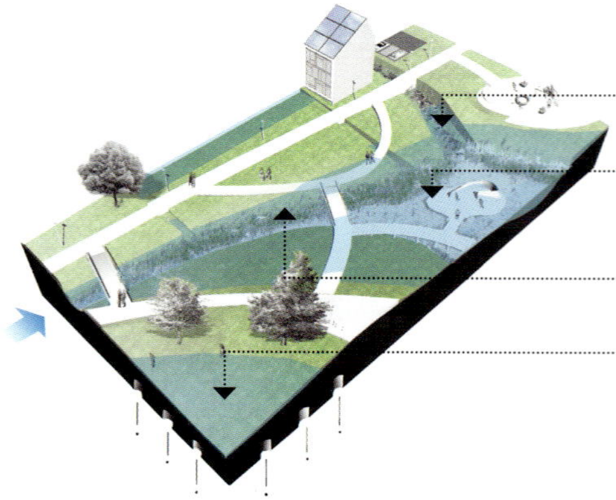

Gravel swales reduce rainwater run off from roads.

Excavated areas may provide dry and wet ponds for wildlife and flood storage.

Planted swales channel water around the site in times of flood.

Park space provides consolidated area for energy infrastructure, such as boreholes and extreme flood storage.

Middle Catchment > Let Rivers Flow

Within the middle catchment of a river, the land is typically less steep than upstream and the rainwater catchment area far greater. Rainwater takes longer to reach the river at this point, but when combined with rainfall collected from across the catchment, it can result in large volumes of water that may remain on land for days following a flood.

Flood volumes are too large to consider storing on site and it is far more important to allow floodwater to pass through the site to avoid causing flooding upstream. Topographical level variations within the development site can help to take standing water away from buildings.

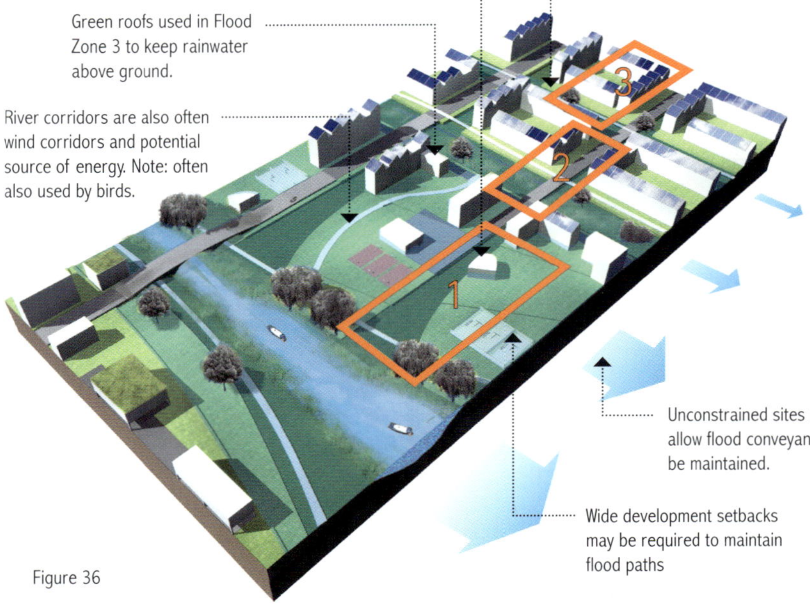

Figure 36

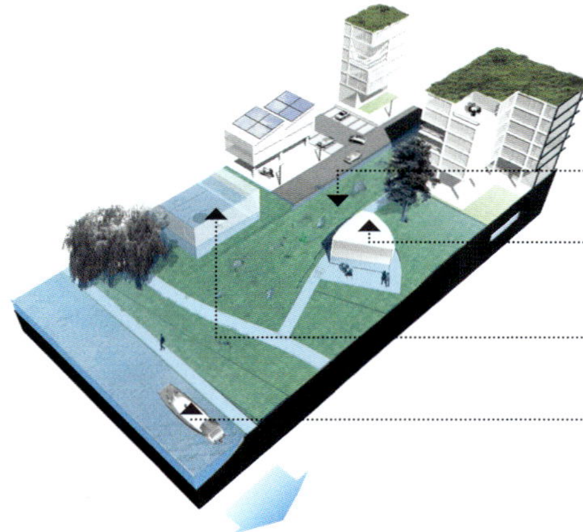

SECTION 1 - Riverside

Generous development set backs from the river may be required.

Buildings closer to the river may need to be resilient to flooding even in lower risk areas.

Hard play surfaces and paths can be cleaned quickly after a flood.

Deeper water bodies can provide opportunity for house boats.

SECTION 2 - Road corridors

Taller buildings with elevated amenity space can reduce development footprint.

Main roads elevated above flood levels provide safe access.

Planted open channels provide urban cooling and drainage routes for floodwater.

Variation in land levels may help to channel water away from homes.

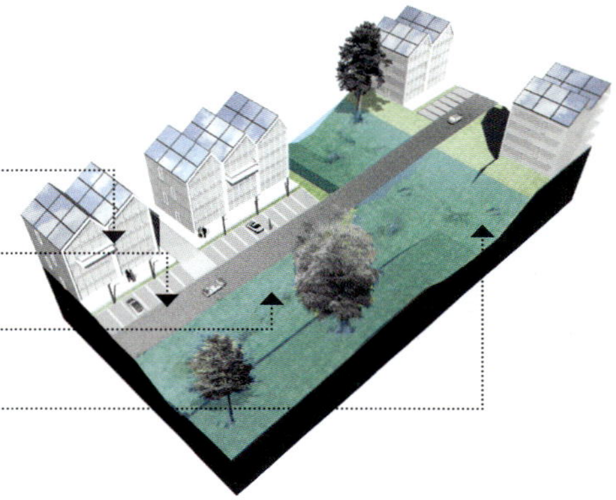

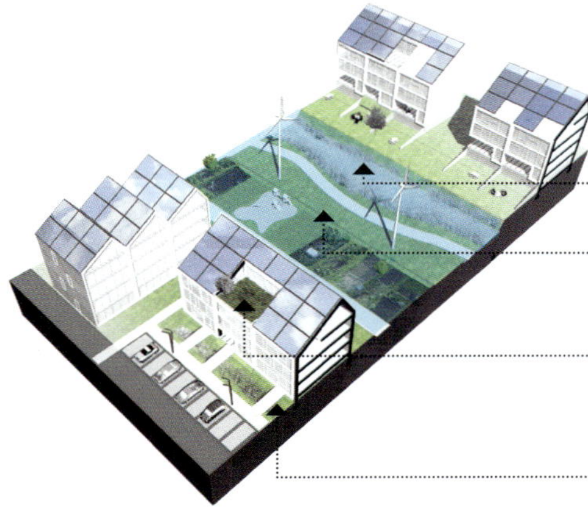

SECTION 3 - Stream corridors

Excavations where the water table is high may create attractive permanent water bodies.

Wide conveyance paths offer multiple daily uses such as allotments, amenity and locations for wind turbines.

Green roofs collect rainwater for harvesting and may be integrated with solar PVs.

Gravel or planted swales attenuate rainwater and make building thresholds.

Lower Catchment > Let Tides Go

Within the estuarine lower catchment of a river the land is often flat and the catchment area typically covers thousands of square kilometres. The river channel is likely to be wide and navigable. The natural floodplain may be embanked and the flood risks not obvious.

Multiple flood sources are likely to exist but the greatest risk is likely to be from the sea. The tide can carry huge volumes of fast moving water which can be very dangerous; but which will typically retreat with the next low tide. Flood volumes are too great to accommodate on development sites. Intelligent spatial planning and structural resilience of buildings can help to manage risks.

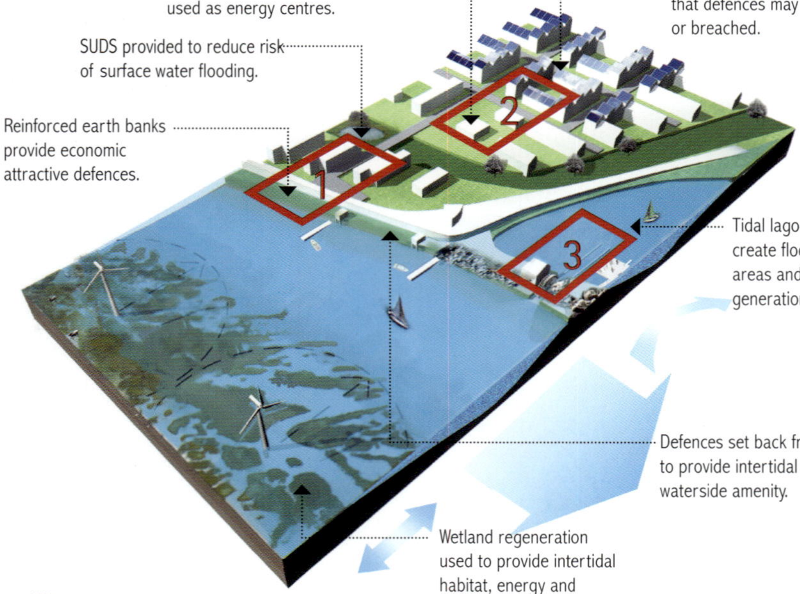

Figure 37

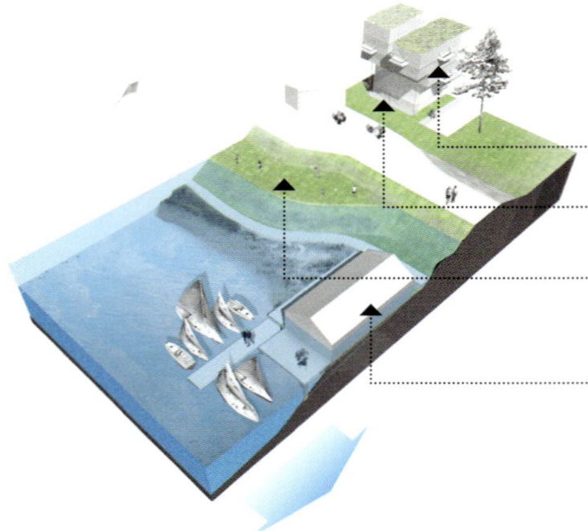

SECTION 1 - Riverside

Buildings set back behind defences.

Lower floors may best be used for 'less vulnerable' purposes such as shops.

When defences are required, wide earth embankments can be used for recreation and habitat.

Buildings adjacent to rivers should be water compatible and not inhabited.

SECTION 2 - Behind defences

Generous elevated balconies or walkways can provide safe access and egress.

Level variation and street layout can slow flood flows, even behind defences.

Concrete frame buildings designed to allow floodwater to pass through help to resist structural collapse.

Rainwater needs to be attenuated and potentially stored, particularly during high tides, when drains can be blocked.

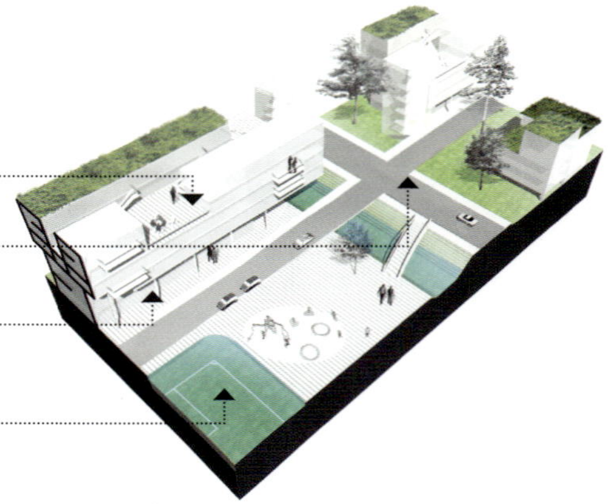

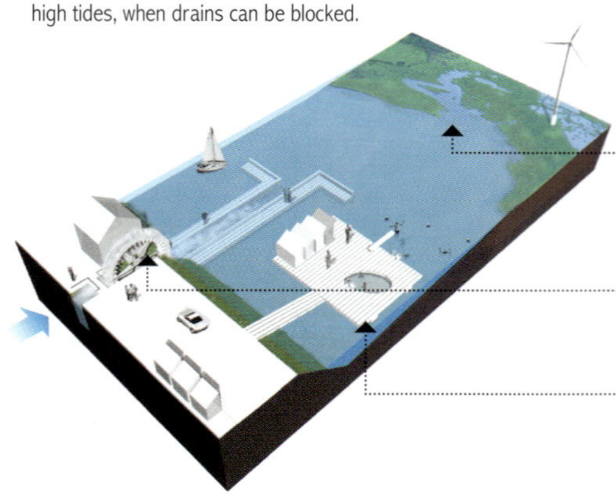

SECTION 3 - Tidal docks

Controlled tidal inundation provides opportunity for habitat such as mudflats and salt marsh and suitable space for large-scale wind power.

Lagoons can be used to store water at high tides, and release it at low tide to drive electric turbines.

Docks or lagoons can provide opportunity for recreation and water compatible development such as bathing houses.

Make space for water, energy & play

Upper catchment > let rain slow

Flood types Fluvial, typically flashy. Also surface water, sewers and groundwater.

Climate change Increased flashiness of flooding from rivers, increased flooding from land and sewers.

Solutions 'Let rain slow' - Slow rainwater run off to reduce pressure on drains and delay rainwater entering the river. Green roofs and underground storage may be suitable in the floodplain. Ground measures, like swales and rain gardens, may also be suitable in the floodplain.

'Let rivers flow' - Allow the river to flow but make room for flood attenuation if possible.

Middle catchment > let rivers flow

Flood types Fluvial flooding dominant. Also surface water, sewers and groundwater flooding.

Climate change Increased flood frequency, extent, depth and duration, increased risk of flooding from land, sewers and groundwater.

Solutions 'Let rivers flow' - Allow the river to flow during a flood by reducing obstacles to floodplain conveyance. Provide paths for floodwater to return to the river afterwards.

'Let rain slow' - Green roofs and underground storage may be suitable in the floodplain. Ground measures, like swales and rain gardens, are suitable in the floodplain.

Lower catchment > let tides go

Flood types Tidal flooding dominant. Also fluvial flooding, surface water, surcharged sewers and groundwater flooding can also be present.

Climate change Increased flood frequency, extent and depth. Increased duration of flooding from land, sewers and groundwater. Increased residual risk behind defences.

Solutions 'Let tides go' - Avoid high tides by letting water pass around the site into dedicated flood storage areas. Build in resilience behind defences and emergency escape routes.

'Let rivers flow' - Allow the river to flow during a flood by reducing obstacles to floodplain conveyance. Provide paths for floodwater to return to the river afterwards.

'Let rain slow' - Attenuate rainwater run off to reduce risk of surcharged sewers or surface water flooding.

Figure 38

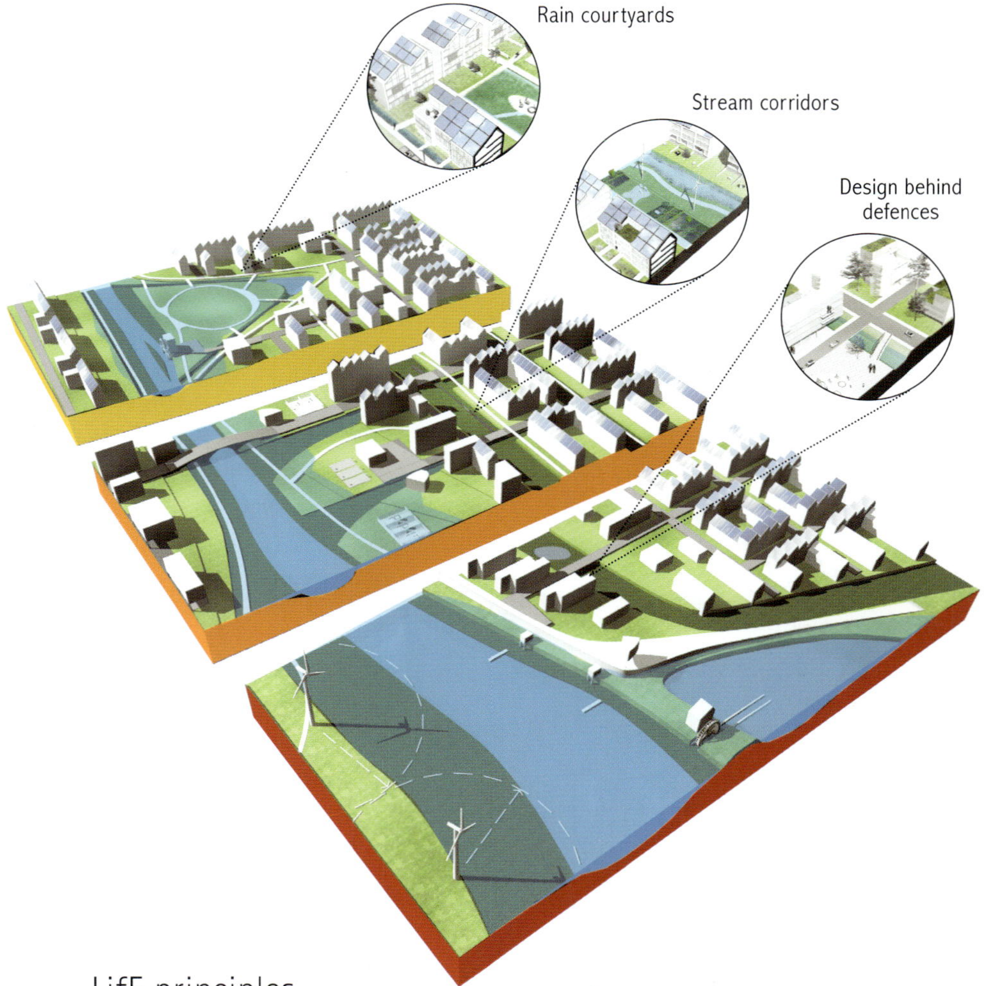

LifE principles

- **Provide space for the river**
- **Provide space for rain**
- **Create space for amenity**
- **Integrate community needs**
- **Design to be adaptable**
- **Reduce car dependency**
- **Provide space for energy**

Conclusions >

The LifE project found that integrated planning can help to achieve zero carbon development more cost effectively, it can create better quality environments to live and work in, it can help manage flood risk and it can help adapt to climate change in the future.

Managing Flood Risk

The case studies demonstrated that responsible development in flood risk areas is possible. They showed that non-defensive, ecological, flood risk management could be incorporated within developments without significant impact on cost (only 1% of the total cost of sites 1 and 2).

The case studies demonstrate various ways in which the means of managing flood risk can be an asset to development, facilitating the integration of renewable energy technologies and providing wider benefits to the community.

The cost of integrating SUDS was not found to be prohibitive and was possible to be provided within the development layout.

Consideration of the wider context, particularly identifying the location of the sites within the river catchment, was essential to understanding the flood risk and the suitability of proposals.

Extreme climate change scenarios and future storm events should be considered when planning in areas at risk of flooding, in particular areas where the impact from climate change would be greatest, such as low lying and coastal land. In each case study capacity for future flood storage or flows was designed into the master plans to ring fence space for adaptation.

Energy

The cost analysis of the three study sites showed that the most significant capital cost increase was a result of the energy technologies.

The LifE project highlighted the difficulty of achieving Zero Carbon through on-site energy technologies alone. This would be easier to achieve if energy networks were developed as part of a larger district.

Multi-functional Land Use

Land used for flood risk management was found to be compatible with renewable energy technologies, therefore helping to make developments zero carbon. The rain gardens and rain corridors in Hackbridge and Peterborough showed that sustainable drainage could be integrated with open space provision and used for recreation.

Multi functional land use was found to be essential to achieving minimum open space standards.

On each of the study sites the demand on the available land for development, river flooding, renewable energy, habitat and recreation was greater than the available land. This is likely to be true when higher density developments are sought unless these demands are combined. Combination of these functions may enable a far higher provision of facilities on any given site.

The case studies for the LifE project show a number of ways in which this can be done without diminishing quality of space and place making. In fact, the proposals showed a number of ways (such as rain gardens, dry ponds, tidal lagoons and energy parks), which could be used to enhance the quality of our towns and cities.

Conclusions > Issues and Opportunities

Assessing Flood Risk

Predicting flooding is complicated. The likelihood of flooding on a site is governed by many factors, natural and man-made. Sometimes it may not be obvious if a site is at risk of flooding or if it is, where it will occur.

It has always been difficult to predict exactly when a flood will occur but it is possible to predict the likelihood that a flood may occur at a given location. The rate with which our environment has changed has made it difficult to keep these predictions up to date, even with sophisticated computer modelling (every new house built or demolished, every road laid, tree felled and drain installed has had an effect on the water cycle).

Climate change has increased uncertainty over future weather patterns and long-term climate trends. This has made it more difficult to predict and map the likelihood of flooding in the future. What we consider to be extreme events now could become commonplace in the future.

It may not always be appropriate to develop on sites at risk of flooding. However, the LifE project has demonstrated that, by working with local stakeholders, the Environment Agency and other experts, it may be possible to manage and reduce unacceptable levels of risk such that sites may be considered for development.

Managing Flood Risk

Managing flood risk requires a response to both the probability of being flooded and the consequences should a flood occur. However, all developments can help to reduce flooding by reducing the rate of rainwater run off through use of SUDS.

Accommodating sufficient floodwater to reduce risk typically requires a lot of land. This is rarely available on individual sites. However, creating space for water on numerous sites may provide overall benefit.

Careful site planning and resilient design should be used to manage the consequences of flooding. Allowing floodwater to flow onto sites in a controlled and predetermined way may help to mitigate the consequences of a flood and maintain continuity of daily life. It may be important to allocate land now that can be adapted to manage flood risk in the future.

Identifying the location of a site within its river (or water) catchment may help to understand flood risk. It can also help in assessing the suitability of proposals and the effect they may have elsewhere.

Sites must not be considered in isolation. Inappropriate development may affect neighbouring sites negatively. Appropriate and intelligent development may benefit neighbouring sites. An approach to managing flood risk should be drawn from beyond the red line boundary of the site. Development frameworks should be regularly updated to align with neighbouring areas.

Meeting Zero Carbon

Realising zero carbon development is challenging. Providing sufficient energy to meet all the power needs on a site requires the use of multiple renewable energy sources. Some sites may have limited resources available. This may increase dependence on biomass to achieve zero carbon. It is important to use the on-site resources efficiently. Orientating buildings to make use of the sun is likely to be essential.

Renewable energy technologies require space. Sites near rivers and coasts often have more sources of renewable energy available. This may help to realise zero carbon development more cost effectively. It may also help to support higher density development.

Areas set aside for flooding (and natural floodplains) may be shared with renewable energy technologies such as GSHPs, wind and tidal turbines.

Renewable energy technologies are expensive. Sites that are windy and wet may currently allow more cost effective energy generation than those that are not.

Reducing heat loss in buildings requires high levels of air-tightness. This may necessitate mechanical ventilation and cooling.

Conclusions > Issues and Opportunities

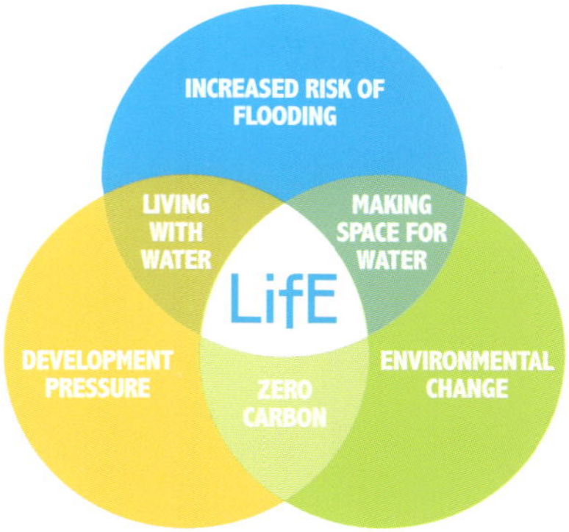

Land Use

Areas of open space are often under-provided on developments where land is more valuable for building.

Land allowed to flood and/or provide for renewable energy may also be used for recreation. When located close to or within development sites this may provide generous open spaces and support healthier environments.

Soft landscaping used to manage flood risk, such as rain gardens and rain corridors, may help to reduce air temperature in and around buildings, mitigating the urban heat island effect. This may also create more attractive areas to live and work and wildlife havens.

Large areas of land or water set aside for flood management may provide habitat, which is being lost to urbanisation and sea level rise.

The opportunity to integrate areas for rainwater, floodwater, energy and recreation within and throughout developments, promotes a greater connection with the natural environment. This approach can be applied to new and existing developments.

LifE Benefits

Adopting the LifE approach may result in higher quality developments, through meeting more objectives, and may help to identify solutions between conflicting parties.

The LifE approach may help to enable sites to be redeveloped that may otherwise be left dormant.

Applying the LifE principles to existing towns may help to identify long-term solutions. Areas may be redesigned to provide flood storage or replace drains with SUDS. Redevelopment, even on small sites, within the context of a LifE plan may help to make areas more resilient to flooding and more adaptable to climate change.

Combining a number of planning applications within one collective environmental and flood risk management strategy may help to realise planning approval for difficult infill sites.

The flood risk management measures used in the case studies, such as the Hackbridge Village blue/green and the Littlehampton tidal lagoons, could not only become attractive assets to the development, they may even become regional attractions.

Following LifE

Further, detailed, modelling may help to assess the potential benefits of the LifE approach with greater certainty.

Development designed on the basis of the LifE approach will need to be piloted through the planning system to demonstrate that it can comply with regulations and to improve the application of the LifE approach in future policy or guidance.

There are few built examples of this kind in the UK. It will only be possible to fully assess the success of LifE developments once they have been piloted and constructed.

There may need to be financial incentives (such as through land provision) to encourage more sustainable development.

Detailed information, such as flood levels and land topography, can be difficult and expensive to obtain. This needs to be made more accessible and easy to understand to help people to make informed decisions. This could be provided through the national planning portal.

Planning policy needs to be evolved to make it easier and quicker to support proposals that incorporate the LifE approach.

Implementing the LifE approach

Understanding issues and objectives from the outset is fundamental. This includes local and national issues and relevant future policy. The objectives of developers should be aligned with those of councils, local residents and stakeholders, such as the Environment Agency.

These issues and objectives should be reviewed in the context of regional and neighbouring policy, particularly as water and river catchments extend across physical and political borders.

Knowing what questions to ask and where to find the answers is essential to establishing a good understanding. The LifE handbook provides an accessible and non-technical introduction to flood-risk and development.

Working with stakeholders from an early stage and addressing their needs may make the planning process easier, the resultant design more successful and should help to optimize development capacity.

True integration of ideas is more likely to occur when cross-disciplinary understanding is developed from first principles. This can lead to innovative and site specific responses that add value to the development, such as those demonstrated in the LifE case studies.

Integrating Non-defensive Flood-risk Management

— Arrange sites so that more vulnerable building uses are located in areas of lowest risk.

— Locate less vulnerable amenity activities (playing fields, car parks, markets...) in flood risk areas.

— Design these areas to be resilient to flooding and able to recover quickly after a flood.

— Design buildings to cope with flooding, and use the most appropriate type of building for the type of flooding.

— Preserve the existing floodplain where possible and use it for amenity for residents if possible.

— Engineered landscape can be used to regulate flood flows, and double as recreation space.

— SUDS may be integrated with play areas, gardens and even roads.

Achieving Zero Carbon and Other Benefits

— Land set aside for flooding may be used to locate renewable energy technologies, in particular for large-scale wind turbines.

— By providing more greenery within the site, urban heat gain can be reduced, reducing the need for artificial cooling.

— Development near existing communities and public transport may reduce the need for goods and people to travel.

— Reducing the quantity of hard paving and roads on developments reduces the carbon used in construction.

— Landscape design measures to manage flood risk may be less carbon intensive than engineered defences.

— Creation of wetland areas can improve habitat provision and environmental gain.

— Non-defensive flood risk management may help to reduce risk elsewhere.

Conclusions > Implementing the LifE approach

How an Integrated Approach can Reduce Costs

- Integrating renewable energy on-site is the most expensive component of the LifE approach. Promoting the use of cost effective technologies is essential to reducing costs.
- After biomass, large-scale wind turbines are currently the most economic technology to provide on-site renewable energy but require large areas of land to be free from development. Floodplain and flood storage areas can provide suitable space.
- Cross platform solutions can provide the opportunity for shared funding or use of alternative revenue streams.
- Non-defensive flood-risk management measures can be cheaper than hard engineered solutions. They may also help to mitigate the costs incurred from flood damage by providing more water-resilient developments.
- Reducing the amount of hard landscaping may save money.
- Omitting conventional underground drainage in favour of complete SUDS systems may save money.

Considerations

- Developing an integrated plan requires more input and information upfront. This can increase start up costs.
- Detailed and costly modelling may be required to demonstrate that it is safe and does not increase risk elsewhere.
- Objections of perceived reduction in standards of defence
- Wind turbines and other perceived unsightly technologies may meet objections.
- Transferring responsibilities (such as flood storage or SUDS adoption by local authorities) can be problematic. Local management structure may also be need for local energy systems.
- The maintenance costs of soft landscaping can be greater than hard landscaping.

What Works Well

Early information gathering helps to build understanding of issues and opportunities.

Understanding the location within the water catchment helps to understand the wider context.

Consideration of future extreme events should be used to inform current decision making.

Coherent land assembly may help to enable the most appropriate master planning at scale.

Organising buildings into compact blocks of development helps to preserve land for other uses such as SUDS or renewables.

Lower density development may allow more land for flood mitigation and make zero carbon easier to achieve. It may also provide greater surface area for solar PVs.

Creating more natural riverside landscapes may help to reduce the extent of flooding.

Well designed SUDS such as those proposed in the case studies (rain gardens, rain corridors, home zones, and rain courts) can integrate well with landscaping.

Public parks and open space can be used for flood alleviation or SUDS.

Safe access to development can be integrated with rain water controls like green roofs, and with raised private amenity.

Linking green spaces create biodiversity corridors as well as drainage paths to help recovery after flooding.

Reducing car parking can help to provide more space for parks and play.

APPENDIX A
Site 1, River Wandle, Hackbridge

Contents

SITE 1 > The River Wandle at Hackbridge

Site Analysis

Formulating the Design Brief

Design Objectives

Review

The River Wandle at Hackbridge

Site Analysis >

Site Analysis

SITE 1 > The River Wandle at Hackbridge

Site 1

Location....................South West London
River..........................Wandle
Catchment.................Upper Catchment
Current use...............Brownfield
Area..........................25 hectares
Units..........................1800
Density......................72 units/hectare

Understanding the Location

Hackbridge is located in South West London (see figure A1) in the London Borough of Sutton, along the edge of the River Wandle and between the A237 and B277. The local population is about 10,000 people.

The local authority is considering plans for between 1500 and 2000 new homes across a number of sites in Hackbridge. Collectively these sites, shown in Figure A2, constitute the LifE study area.

The target number of homes being considered by the local authority reflects the targets set out in the Mayor's London Plan 2004, for mixed terraced houses and flats of between 50 and 110 units per hectare. The total area of all of the development sites is 25 hectares.

All of the sites are within a 10 minute walking distance of the local railway station, Hackbridge, which provides connection into central London with a journey time to Victoria Station of 25 minutes. There are also a number of regular local bus services.

Figures 3a to 3g show photographs of the Felnex and railside sites and the surrounding area.

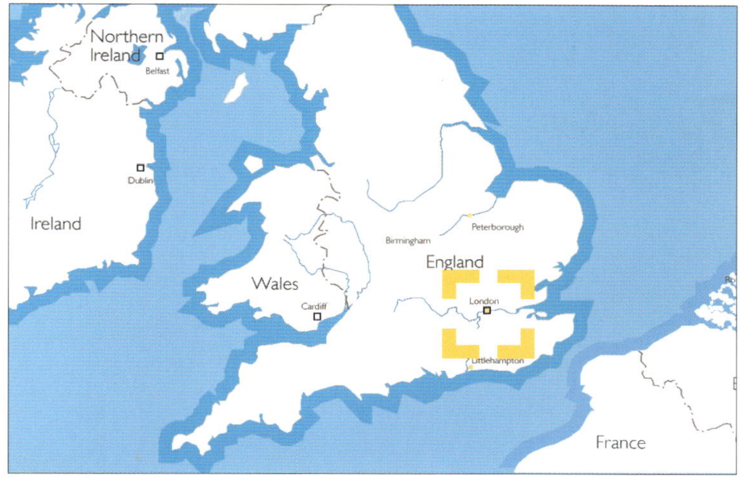

Figure A1 - Location plan

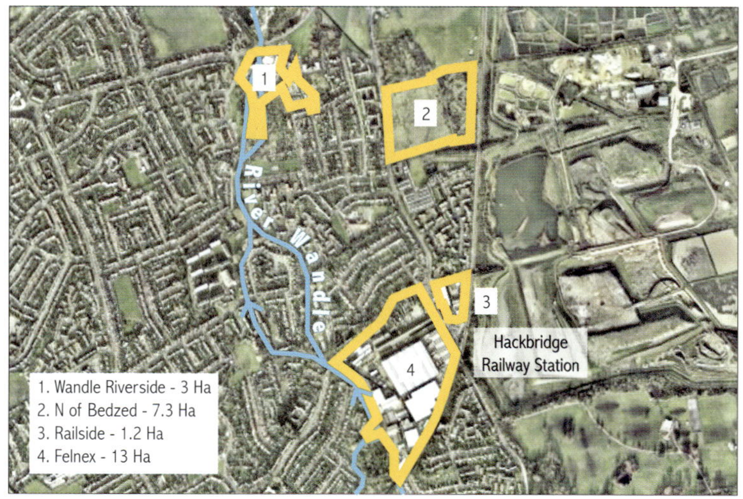

Figure A2 - The site areas

Site Photographs

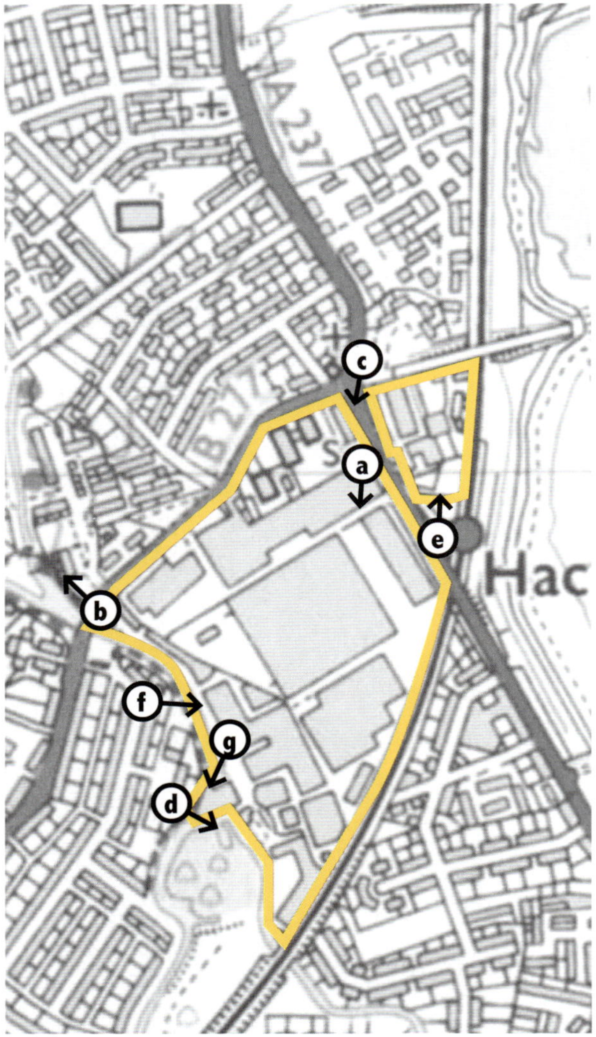

Figure A3 - Key to photographs

Figure A3a - View of Hackbridge railway station and entrance to the site.

Figure A3b - View from the road bridge

Figure A3c - View towards the corner of the site, showing the local centre shopping parade

Figure A3d - Weir adjacent to former mill

Figure A3e - View of Hackbridge station

Figure A3f - Offices on the east bank of the Wandle

Figure A3g - Culverts under the mill site

Prior to the industrial revolution Hackbridge was a crossing point over the River Wandle surrounded by farmland.

Prior to the 17th century three flour mills were built along the river. During the 17th and 18th centuries the number of mills rose to 41 as new industry was established along the river for leather, copper, gunpowder, snuff, paper, linseed oil and flour. Some reports suggest that this figure rose further to 94 in Victorian times.

Figure A4 indicates the locations of three former mill sites on historical maps of Hackbridge. These are the Hackbridge mills, the Culvers Bleaching Grounds and Corn mills, and the Leather mill at Goat Bridge.

Since the arrival of the railway line in 1865 the area has progressively turned from farming into a residential suburb of London. In recent years there has been a large turnover of property in Hackbridge as first-time buyers purchase affordable homes before moving on to more affluent nearby areas.

Residential buildings have now been developed along much of the River Wandle at Hackbridge, however, a large number of industrial buildings still occupy riverside locations.

The River Wandle has a high heritage value with numerous historic estates, archaeological sites and industrial history along its length.

Figure A4 - 1865 Ordnance Survey map showing former mill sites

The River Wandle

The River Wandle is a tributary to the River Thames. It flows from its source a few kilometres south of Hackbridge at Waddon and Carshalton to the Thames at Wandsworth.

The Wandle is a fast-flowing, low altitude watercourse with small catchment (13.44 km^2) and calcareous geology, considered to be one of the finest chalk streams in London.

Hackbridge is located within the upper catchment of the river, approximately 25 m above Ordnance Datum Newlyn (ODN). The river flows from south to north through the site across a shallow landfall, shown in figure A7.

The river retains natural banks for much of its length, but other areas remain heavily managed, with culverts, artificial channels, run off ditches and subterranean stretches (see figures A5 and A6).

Figure A5 - Canalised section of the Wandle

Figure A6 - Canalised section of the Wandle

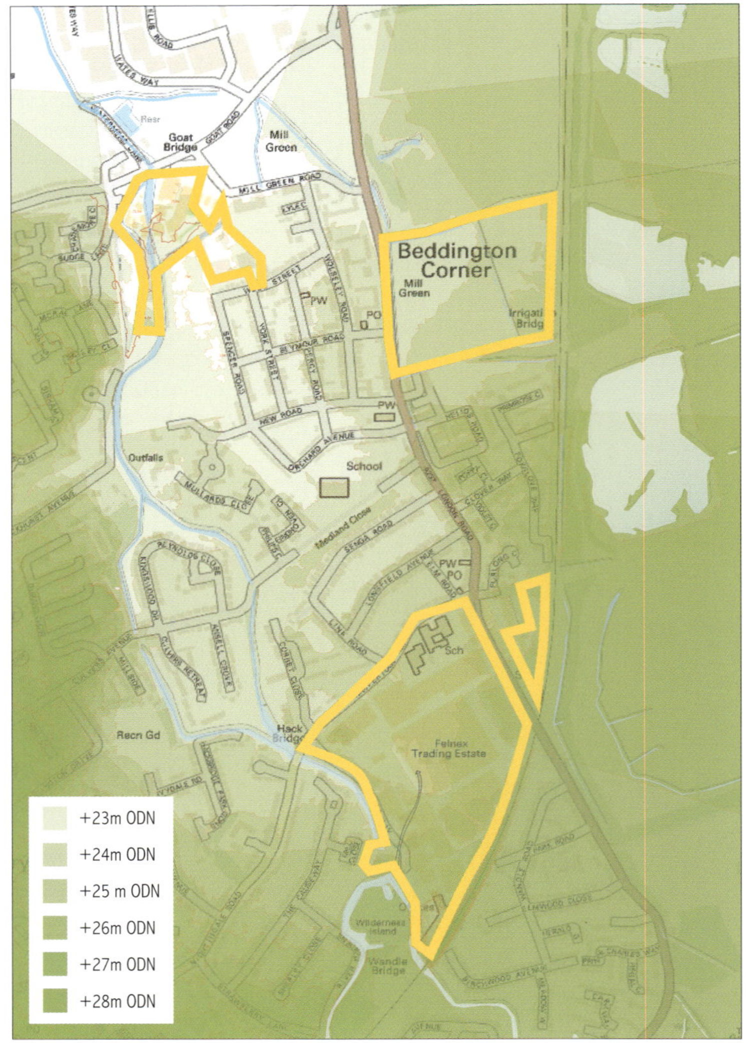

Figure A7 - Topographical plan

The upper River Wandle is identified as a Site of Metropolitan Importance for Nature Conservation. The local landscape character is shown in figure A9.

Three Local Nature Reserves are located within the vicinity of the sites:

Wilderness Island is a 2.72 ha wetland area in the centre of the river, to the south of the Felnex Site. It contains a variety of habitats including a mosaic of ponds, sedge bed and damp hollows, supporting 30 species of birds and 70 kinds of wildflower.

Wandle Valley Wetland is a 0.62 ha site to the east of the Wandle Riverside Site that includes open water, marginal vegetation and seasonal pools supporting invertebrates such as dragonflies and damsel flies.

Spencer Road Wetlands is a 1.04 ha site to the south of the Wandle Riverside Site and home to the London Wildlife Trust's water vole project, which aims to reintroduce the water vole to the Wandle by 2009.

The Wandle Trail, shown in figure A8, is a 20 km / 13 mile long trail following the course of the River Wandle from East Croydon to the River Thames at Wandsworth showcasing the heritage and wildlife value of the river corridor. The trail includes a number of footpaths and cycle tracks, parts of which are included in National Cycle Network Route 22. The Wandle Trail Arts Programme provides numbered gateways, viewing platforms and distinctive way markers.

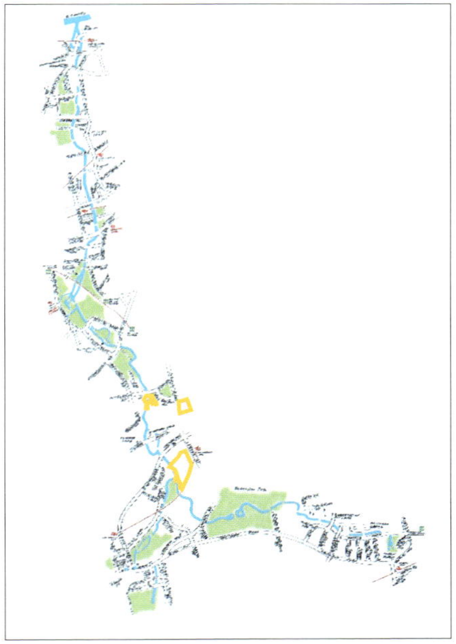

Figure A8 - The Wandle trail

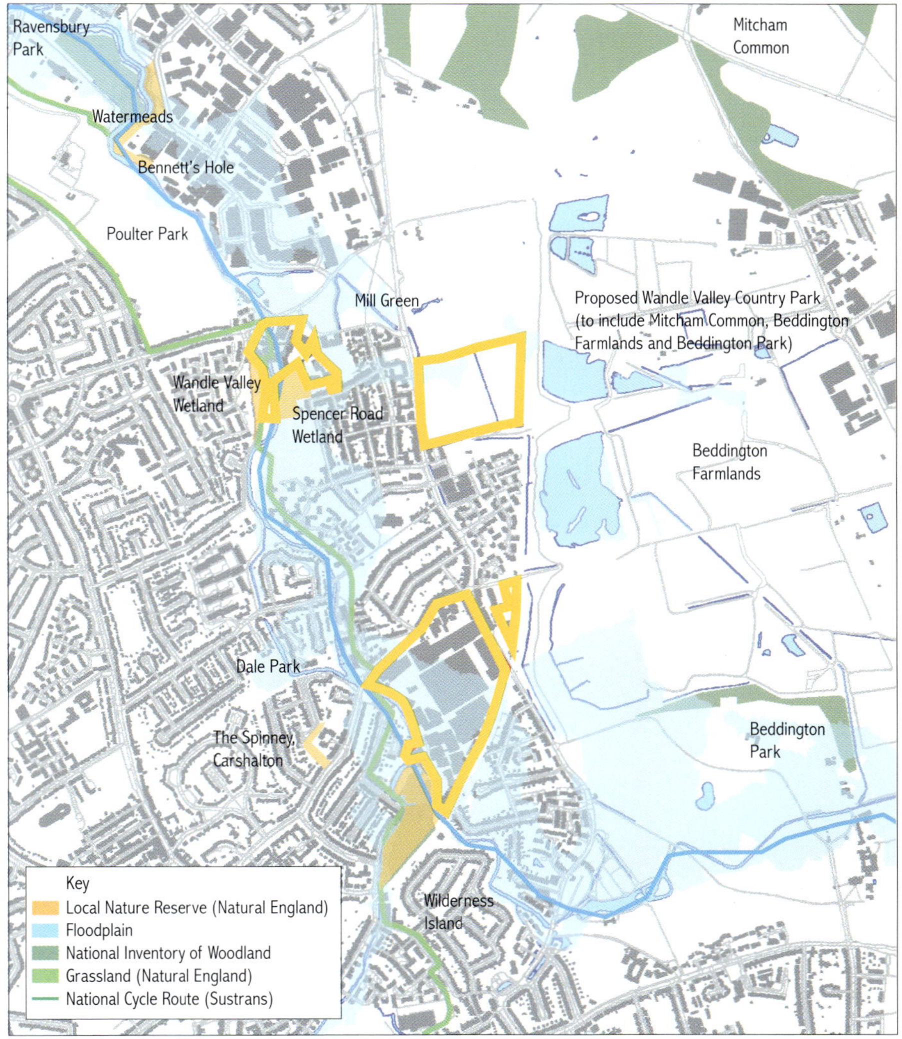

Figure A9 - Local landscape features

Flood Risk

Flood risk in Hackbridge can be from either fluvial flooding (flooding from rivers) or surface water run off (flooding from land and sewers).

The widespread development of the river catchment has resulted in the coverage of large areas of impermeable surfaces, which has reduced the capacity of the ground to infiltrate rainwater. This causes the water to run off the land quickly (causing flooding from land), overwhelming the drains (causing flooding from sewers) and overwhelming the river channel (causing flooding from rivers). Groundwater flooding has not been reported in this area.

Fluvial flood risk is typically of short duration with rapid rise and fall of river levels, typical of this type of catchment. The depth of the floodwater, speed and duration are not likely to be great but the lag period between the rainfall and a flood will be relatively short. This can mean that a flood can still be disruptive, depending on the time of occurrence.

Extent

The Environment Agency flood risk map, shown in figure A10, indicates a 1% probability (100 year) fluvial flood risk area that extends several hundred metres either side of the river course, narrowing on occasion. The 0.1% probability (1000 year) fluvial flood risk area is shown to closely match the 1% extent, only marginally extending the area in a few locations.

The paucity of flood data and other factors (database, detail and quality of the modelling, etc.) has resulted in a flood map that is unreliable. As a consequence, the EA were taking steps (at the time of the project work) to upgrade the modelling and mapping. This modelling was made available for the LifE project. The results suggest that the extent of the 1% flood area, Flood Zone 3, is very narrow through the study area. The 0.1% flood area, Flood Zone 2, is similar to the EA flood map (see figure A12).

Depth

Information on flood depths and durations is very limited. From what is available it appears the 1% (100 year) flood flow increases from about 6 m^3/s at the upstream limit of the site to about 8.5 m^3/s at the downstream limit.

The information available shows an increase in water level of only 0.1 m to 0.2 m between the 5% (20 year) and 1% (100 year) floods at any model node. Extrapolating this information for the 0.1% (1000 year) flood suggests an additional increase in water level of the same order.

From the contour information available extreme depths were assessed to be less than 0.5 m for the 1% (100 year) flood.

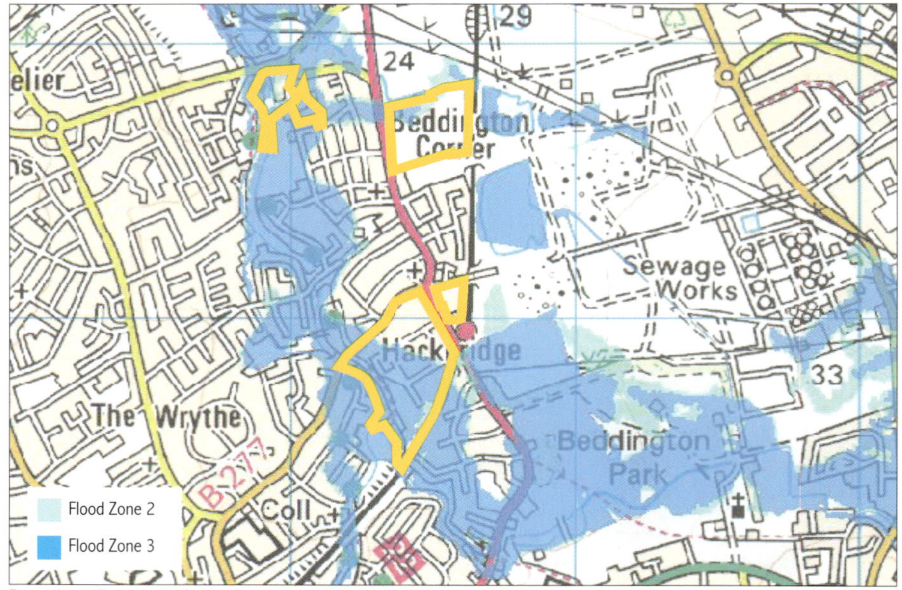

Figure A10 - Environment Agency flood risk map

Flood Zone 1 (Low Probability)
Areas with less than 1 in 1000 year (0.1%) annual probability of flooding.

Flood Zone 2 (Medium Probability)
Areas with between 1 in 1000 year (0.1%) and 1 in 100* year (1%) probability of flooding.
* 1 in 200 year (0.5%) probability of flooding from the sea

Flood Zone 3a (High Probability)
Areas with between 1 in 100* year (1%) and 1 in 20 year (5%) probability of flooding.
* 1 in 200 year (0.5%) probability of flooding from the sea

Flood Zone 3b** (Functional Floodplain)
Areas with greater than 1 in 20 year (5%) probability of flooding.
** This is not shown on the Environment Agency website.

Duration

Stage/time flood hydrographs suggest that the duration for the 1% flood on the north and south sites are approximately 4 hours and 10 hours respectively (see figure A11). However, the flows flood hydrograph indicates that the time period could be less. This uncertainty may reflect the accuracy of the model in assessing the bank height.

Flood Risk

Flood risk is described in terms of the sources of flooding, the pathway or route by which the flood would occur and the receptor such as homes or farmland; the hazards for the site are set out in Table A1.

There is a great deal of uncertainty about the current flood risk on the Wandle due to lack of empirical information and inaccuracy of current models. Therefore the EA has recommended a precautionary approach to assessment of flood risk on the River Wandle. This is to add 400 mm to flood levels predicted by the existing (old) model.

Surface Water Flooding

The EA does not currently provide flood maps for surface water, groundwater or sewer flooding.

No evidence of flooding from the land or sewers was provided, however, as a precautionary approach it was assumed that flooding could potentially occur.

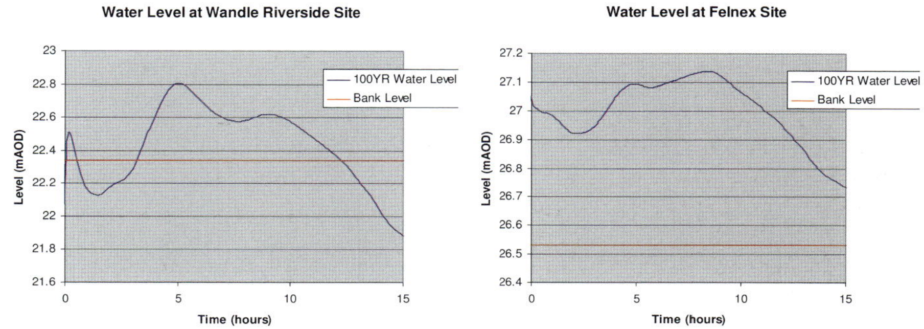

Figure A11 - Flood hydrograph

Site 1, Flood Risk		
Sources of Flooding	Pathway	Receptor
Flooding from Rivers	Overtopping of river banks	Homes
Flooding from Land	Run off from the roads and land	Businesses
Flooding from Sewers	Overflow from drains	Light industrial units

Table A1 - Flood risk to the site at Hackbridge

Effects of Climate Change

Climate Change

In the future, climate change is expected to result in a lower mean annual rainfall, however, the intensity of rain storms is expected to rise, leading to increased rainfall events.

Increased River Flows

Small changes to the nature of the catchment, such as increased development, can significantly affect the extent of flooding. The future extent had not been modelled and the EA recommended using the 0.1% probability (1000 year) event as the extent of the future 1% probability (100 year) event. This is shown in figure A13. There were no current recommendations given for the future 0.1% extent.

Information on flood flows and depths was limited, due to lack of empirical data, however, from the information available the increase in water levels were shown as - 0.1 m to 0.2 m between the 5% (20 year) and 1% (100 year) floods at any model node. Extrapolating this information for the 0.1% (1000 year) flood suggested an additional increase in water level of the same order, 0.1 m to 0.2 m.

Figure A12 - Flood risk map current (based on draft modelling provided in 2008)*

Figure A13 - Flood risk map future

* Since this work was carried out the SFRA level 1[31] was published with updates showing a more extensive flood zone 3, similar to that indicated in figure A10.

Available Resources

Energy Sources

The available natural resources that would be suitable for energy generation at Hackbridge are wind, solar, and water.

The indication of mills on historic maps, supported by the presence of existing weirs and mill runs, suggest that hydro energy generation may be possible.

The land area is too small to grow biomass on site, however, fuel could be provided from an off-site supplier.

This site is in a dense urban location and therefore use of wind turbines was not considered viable.

Therefore the following technologies were identified as viable on this site:

- Solar (Photovoltaic and Solar Hot Water)
- Biomass
- Ground Source Heat Pumps
- Micro Hydro

Rainfall

The total area of the site is 250,000 m^2 and the average annual rainfall is approximately 650 mm per year[40]. Therefore the total volume of rain falling on the site is approximately 162,500 m^3 per year.

The 130,000 m^2 Felnex site would receive an average annual volume of rain of approximately 84,500 m^3.

Formulating the Design Brief >

Development Mix

Following guidance and advice from the local planning department an accommodation schedule was established. Using this schedule, energy demand, water demand and amenity standards could be determined.

A conceptual mixed use development of 1800 new homes was proposed across the four sites in Hackbridge. This would equate to a density of 72 units per hectare.

A mix of 800 flats and houses were proposed for the 13 hectare Felnex site, with a further 10,000 m^2 mixed non-domestic buildings, indicated in Table A2. The domestic property density would equate to 62 units per hectare.

Amenity Required

Based on the accommodation schedule the number of inhabitants for the Felnex site was estimated to be in the order of 2080*.

An assessment of applicable national and local planning policy guidance was undertaken to determine likely minimum standards for open space which would be applicable for the Felnex site.

This included a review of the Six Acre Standard (National Playing Field Association, 2001) and Planning Obligations Supplementary Planning Document (London Borough of Sutton, Adopted 22 March 2007).

The requirements for open space for 2080 people were:

- Sports provision – 3.59 Ha
- Outdoor (public) open space – 6.65 Ha
- Private amenity space – 4.22 Ha
- Total required outdoor space – 14.46 Ha

Figure A14 illustrates the areas of outdoor open space required in more detail and the site area available.

* based on an occupancy of 60% of bed spaces

Site 1, Accommodation Schedule

Residential	GIFA (in m^2)	No. of Homes	Total Area (inc circulation)	Parking
1 bed studio	37	40	1,702	52
1 bed flats	50	160	9,200	208
2 bed flats	70	160	12,880	208
3 bed flats	90	160	16,560	320
3 bed houses	100	160	16,000	320
4 bed houses	120	80	9,600	160
5 bed houses	140	40	5,600	80
6 bed houses	160	0	0	0
Residential GIFA (in m^2)		800	71,542	1,296
Commercial			GIFA (in m^2)	Parking
Office Area			1,950	98
Retail Area			5,000	250
Industrial Area			500	25
Educational Area (inc nurseries, creche..)			2,000	100
Health and Community Facilities			500	25
Commercial GIFA (in m^2)			9,950	498
TOTAL GIFA			81,492	1,794

Table A2 - Accommodation schedule

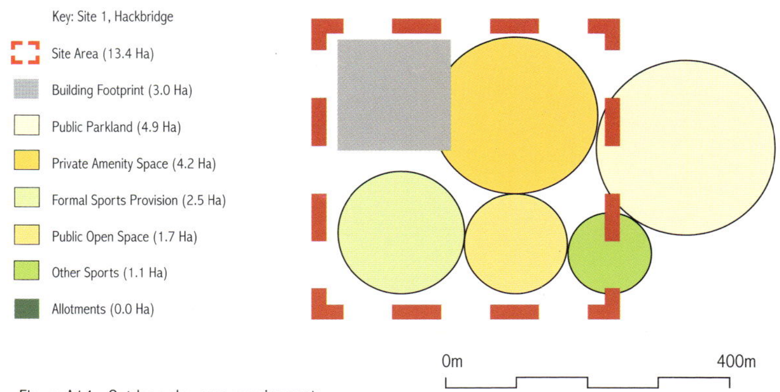

Figure A14 - Outdoor play area requirements

Resources

Energy Demand

Using a range of typical energy consumption figures for different building types (details included in appendix D) the accommodation schedule was used to calculate the total energy demand for the Felnex site, indicated in Table A3 and Figure A15.

Water Demand

The total water demand was calculated as 228,800 litres per day, 89,300 m^3 per year (based on 110 l/p/day for residential, which incorporates dual flush WCs and low flow taps + 11 l/p/yr for commercial).

This water demand is approximately 94% of the total rainfall across the site.

Site 1, Energy Demand						
Energy Demand (MWh/yr)	Space heating	Space cooling	Hot water	Electricity (excl cooling)	Total (MWh/yr)	Percentage
Total Residential Demand	995	0	2,190	2,760	5,950	75.0%
Total Commercial Demand	245	450	135	1,150	1,980	25.0%
Total Energy Demand	1,240	450	2,325	3,910	7,930	100.0%

Table A3 - Energy demand

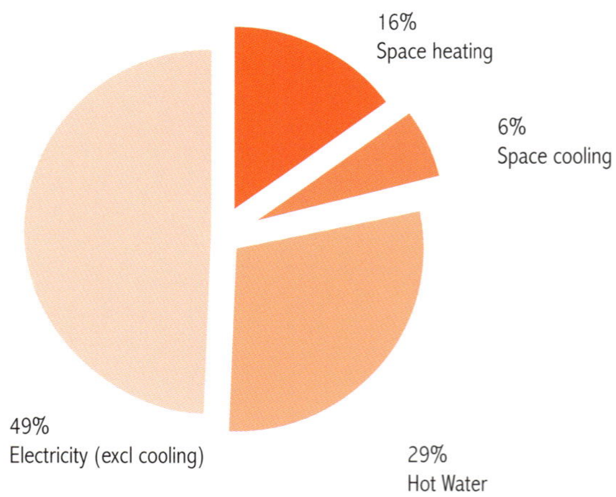

Figure A15 - Energy demand mix

SUDS Requirement

The design standard for surface water drainage is 1 in 30 years, ie that the capacity of the drains can cope with a severe rainstorm, only likely once every 30 years. However, the design standard for flooding is 1 in 100 years. During a 1 in 100 year rainstorm drains are likely to be overwhelmed.

Current EA policy is to aim for 'greenfield' equivalent infiltration and run off where ever possible.

The design standard adopted by the LifE project is to provide surface water drainage that can cope with a 1 in 100 year event plus a 30% increase in peak rainfall to account for climate change to 2115.

The sub-strata for the north site has a high permeability therefore reducing the typical run off rates of between 3 and 5 l/s/ha to 1.6 l/s/ha. However, because both the main riverside sites are former industrial sites it is likely that there will be some contamination and therefore infiltration will not be an option. Additionally no assessment of the groundwater table was made. Therefore an assumption of 5 l/s/ha for the 100 year rainstorm has been used.

Due to the potential for contaminants from the former industrial uses, a sealed system, without infiltration, may be required. Source control and measures such as green roofs may also be appropriate.

A precautionary assessment for the volume of attenuation would be 300 m^3 rainfall per hectare*. This would equate to a total attenuation volume of 3,900 m^3 across the Felnex site. With an average swale depth of 0.5 m an area of 7,800 m^2 attenuation would be required.

The total optimum water requirements are shown in figure A16.

* 300 m^3 per hectare equates to 8 l/s/ha for a 10 hour rainstorm or 5 l/s/ha for a 16 hour rainstorm.

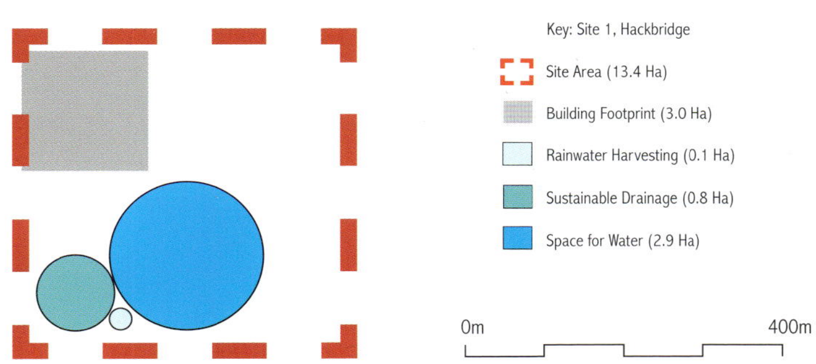

Figure A16 - Spatial requirements for rainwater, attenuation, harvesting etc..

Design Objectives >

The objective was to establish a plan that would help to facilitate the development of each site through shared flood risk management and logical assignment of use.

The strategy was to create resilient design layouts to allow surface water through the site at the same time as creating functional attractive amenity.

Design Objectives

Based on the resources available, land availability in Flood Zone 1 and 2, and the requirement for amenity space, the following strategy was established to enable a suitable development to come forward.

Main Objective

Create a landscape-rich development that provides for the requirements of a medium density suburban population and simultaneously reduces surface water run off and flood risk to lives and property.

Landscape Approach

Develop a high quality landscape for sustainable suburban living that would provide multi-functional open spaces which adapt for flood mitigation, sports and play, biodiversity enhancements, local food and energy. Create new recreation and amenity space along the river, which could also provide attenuation during flood events and environmental gain.

Create versatile land uses between development that can cope with unexpected fast inundation flooding and slow the passage back into the main river course to reduce risk downstream.

Energy and Water Approach

Use a community-wide system to reduce energy loss and enable use of combined heat and power. Top up main energy supply system with utilisation of all available renewable resources.

Aim to provide at least 50% of water requirements for flushing WCs and irrigation from rainwater harvesting to reduce the water consumption to 80 litres per person per day.

Flood Risk Management Approach

The flood risk management approach for Hackbridge is to use non-defensive measures to slow and store floodwater on the larger (Felnex) site to reduce risk to other sites, including the Riverside site, downstream.

And to manage all surface water run off for a 1 in 100 year storm event using sustainable drainage.

Concept Ideas

The concept plan shown in figure A18 shows development arranged around a central park space that connects the train station with the river. The entrance to the site and the road junction were seen as major gateways, requiring higher quality building designs. The main concepts of the plan were:

- Prioritise higher density mixed use development close to the station on higher ground, in Flood Zone 1.
- Arrange buildings on northern site parallel with river channel to allow flood flows through the site (floodplain conveyance).
- Introduce car club to reduce requirement for parking provision and hard landscaping to increase space for amenity and flooding.
- Use pathways through the site to provide new connections to nearby parks, rail and bus links and improve access to the river.
- Improve connections with existing network of footpaths and cycleways, including the Wandle Trail, and enhance with native planting to extend wildlife corridors.
- Create a multi-use open play area that is easily accessible by local inhabitants, for recreation, habitat, flood attenuation, connections and views.
- Use existing water infrastructure, such as mill runs to provide modern energy generation.

Four options, shown in figure A17 were considered for the Felnex site. The 'village blue/green' was thought to provide the widest benefits to the development and the local area by creating a clear sense of place and local identity, improving connections with the local transport facilities, and opening links to the river and wildlife corridors.

The development of the Wandle Riverside site to the north is likely to be more ad hoc, reflecting the local topography, flood paths and defences.

'Court Development' development clusters

'Village Blue/Green' central amenity

'Room around Green' distributed amenity

'Room for the River' riverside amenity

Figure A17 - Plan options for the Felnex site

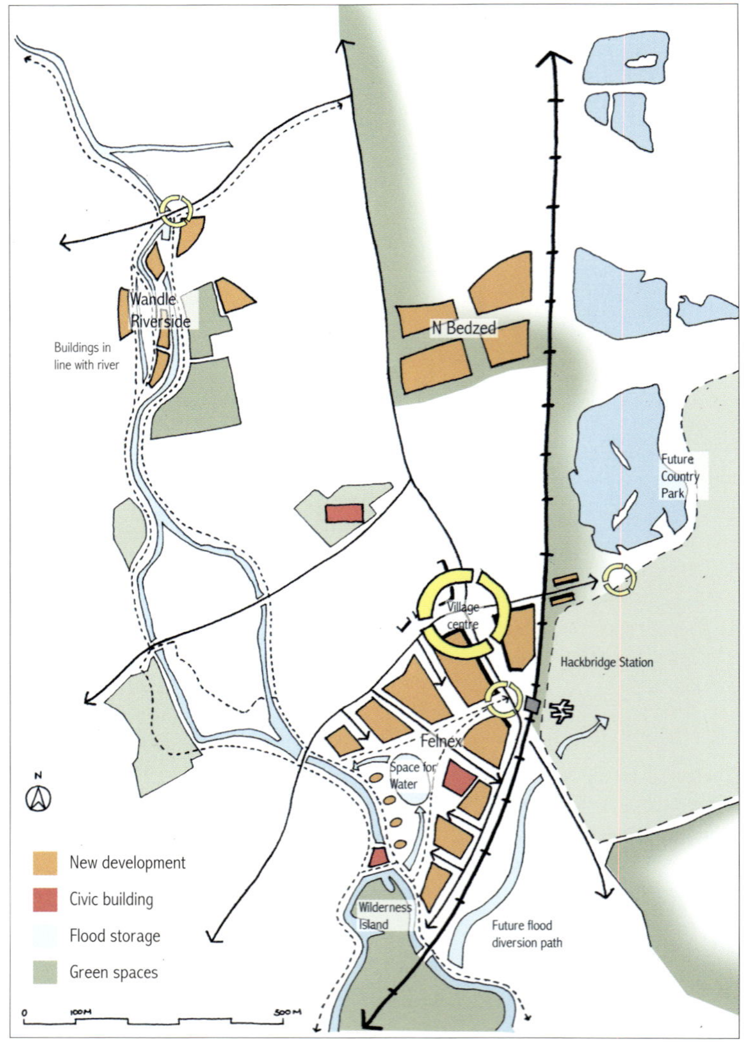

Figure A18 - Concept plan

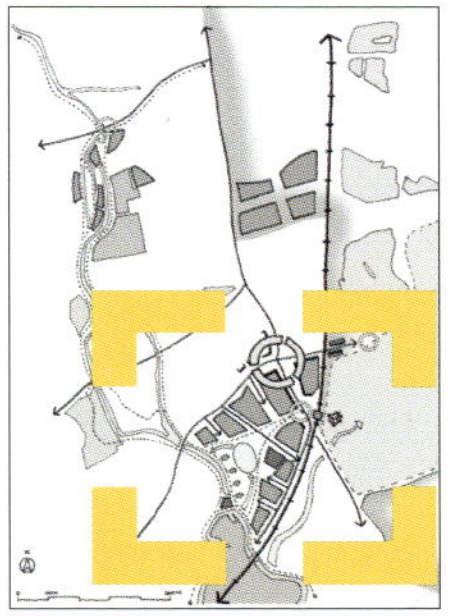

A conceptual masterplan was developed for the Felnex site based on the 'village blue/green'. This is illustrated in figure A19.

Village Green – a flexible informal recreation area with vegetated banks and areas of hard landscaping. The green also provides future flood storage potential and is used to locate an array of boreholes for ground source heat pumps. Mowing regimes would allow for areas of grass at different heights, supporting wildflowers and providing foraging and habitat for birds and insects.

Village Blue – a planted gulley carries water from the river into the 'village blue' at times of flood. This landscaped hollow is designed to regularly accommodate floodwater, which slows the flow downstream, and replenishes a mix of dry and wetland habitats from ponds to reed beds and wet woodland; providing biodiversity for the local school and residents.

Rain Play Gardens – developments are organised around intimate planted courtyards, private gardens and local areas of play. Play areas are surrounded by sunken gardens that provide rainwater attenuation that feeds into communal underground greywater systems. The planting creates a high quality buffer to the play area, without obscuring site lines (for security) from the houses and flats.

Figure A19 - Conceptual masterplan for Hackbridge Felnex and rail sites

Village Centre – a mixed-use urban gathering area connects the station, local shops, the school and the village green. Multi-purpose market squares will accommodate regular and one-off events, and routine activities such as drop-off and collection of school children. A sunken ball court/MUGA doubles as an attenuation area for surface run off.

Energy School – the new school is located above the flood level but centrally within the development and linked to the neighbourhood energy centre. The school provides the ideal location for community management of the energy system and supply chains; and local management and education of flood risk alleviation measures.

Threshold Swales – car parking and paths are separated from the buildings by gravel or planted swales with gentle ramps rising to the homes and creating a defensible threshold.

Riverside Room – a riverside walk combining gravel paths and areas of boardwalk extends the Wandle Trail through a naturalised wetland riverbank. Excavations create low-lying land adjacent to the river for stands of marginal vegetation and pockets of wet woodland.

Energy and Resources

Due to the limited range of renewable resources available on site, biomass was required to provide the majority of the total energy demand.

To provide the additional electricity requirement (3,015 MWh/yr) 30,150 m^2 of solar photovoltaic panels would be required+. However, the available roof area of the whole development, after allowing for plant, access and north facing areas, was found to be 21,700 m^2 ++. When this area was reduced further to allow for 1,100 m^2 of solar hot water panels the electricity deficit was found to be approximately 1,055 MWh/yr.

The energy technologies in order of contribution were:

— Biomass CHP (heating, hotwater and electricity)

— Solar PVs (electric)

— Off-site Renewable Electricity

— Ground Source Heat Pumps

— Biomass Boilers

— Solar Thermal (hot water)

To achieve Zero Carbon on this site the energy shortfall would need to be met by accredited off-site renewable electricity generation.

The proposed energy solution is shown in Table A4.

Technology	Yearly energy demand met *	Which demands met	Further info	% of total
Biomass CHP	1,850 MWh 500 MWh 1,065 MWh	HOT WATER: 100% of flats & commercial; 50% of 100% of houses SPACE HEATING: 50% of residential ELECTRICITY: contribution	Approx 200kWe biomass CHP required. Annual fuel use approx 5,300m³ wood chip, or 1,300m³ wood pellet **	42.0%
Biomass boilers	545 MWh	SPACE HEATING: 50% of residential; 20% of commercial	Annual fuel use approx 800m³ wood chip, or 200m³ wood pellet **	6.7%
Solar water heating* **	455 MWh	HOT WATER: 50% of 100% of houses	Approx 1,000m² total active panel area required	5.6%
Ground source heat pumps (heating)	190 MWh	SPACE HEATING: 80% of commercial	Sizing for peak demand is inefficient, therefore 80% of yearly demand only.	2.4%
Ground source heat pumps (cooling)	390 MWh	COOLING: 80% of total	Sizing for peak demand is inefficient, therefore 80% of yearly demand only. Approx 110, 70m deep boreholes required (heating & cooling)	4.8%
Micro hydro	20 MWh	ELECTRICITY: Small contribution	Estimated yearly output of micro-hydro installation at Leather Mill	0.2%
PV (required)	3,015 MWh	ELECTRICITY: To make up balance of demand; to run ground source heat pumps; to run chiller to satisfy remainder of cooling demand	Approx 30,150m² total active panel area required. This will not fit on the development ****	38.3%
PV (used)	1,960 MWh	ELECTRICITY: To run ground source heat pumps; to run chiller to satisfy remainder of cooling demand; elec output from largest PV installation possible onsite	Approx 19,620m² active panel area, which is max that will fit on-site ****	25.3%
Offsite electricity	1,055 MWh	ELECTRICITY: To make up balance of demand	The use of wind turbines has been ruled out for the site, and therefore use of offsite renewable electricity is required to make up shortfall to achieve 'zero carbon'.	13.0%

Table A4 - Proposed energy solution

\+ It would be possible to burn more biomass to produce the extra electricity, however it would generate additional 'waste' heat, which would increase surrounding air temperatures, with the possibility of increasing the need for mechanical cooling, especially during summer.

++ If the building heights were reduced to a maximum of 3 storeys and all of the roofs angled exclusively south the development footprint would increase by approximately 60% and therefore the roof area should be sufficient for the solar PV requirement.

* Note that these figures have been rounded to the nearest 5 mWh.

** For total biomass demand, area of willow coppice required to be self sufficient in biomass = 71 hectares (or 0.09 hectares per residential property).

*** Flat plate panels assumed

****The total roof area for the development is around 31,000 m². Not all of this area will be available for solar panels (solar water heating and PV). Assuming 70% would be suitable for panels (this may also include some suitably orientated façade areas), total available area = 21,700 m². Assuming 1,100 m² roof area required for solar hot water (allowing 10% for frame etc), this leaves 20,600 m² for PV, equivalent to around 19,620 m² active PV panel area (assuming 5% for frame etc).

Figure A20 illustrates the proportional contribution of each energy technology. Figure A21 indicates the spatial demands of the energy technologies, including the amount of land needed to grow biomass to support the annual fuel demand.

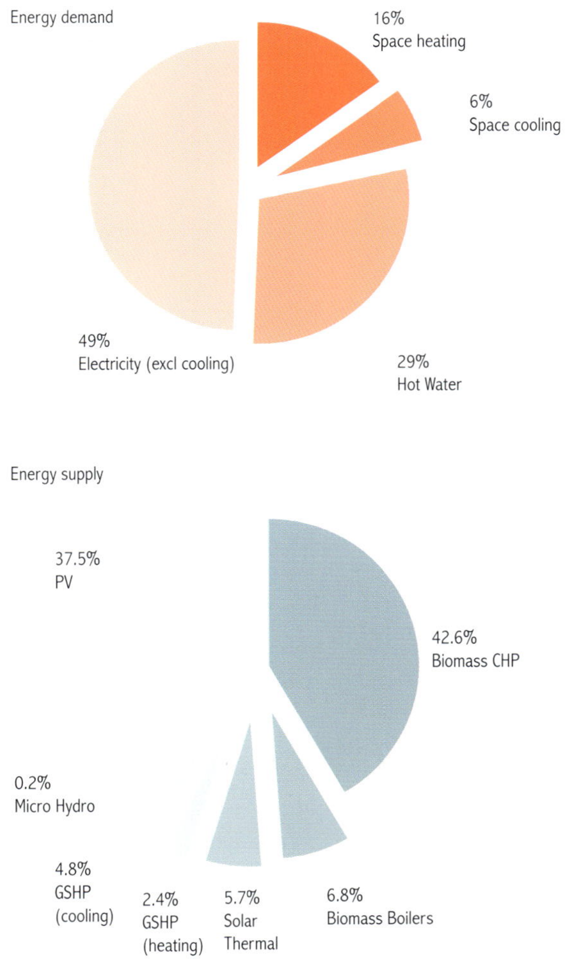

Figure A20 - Energy technology contribution to total demand

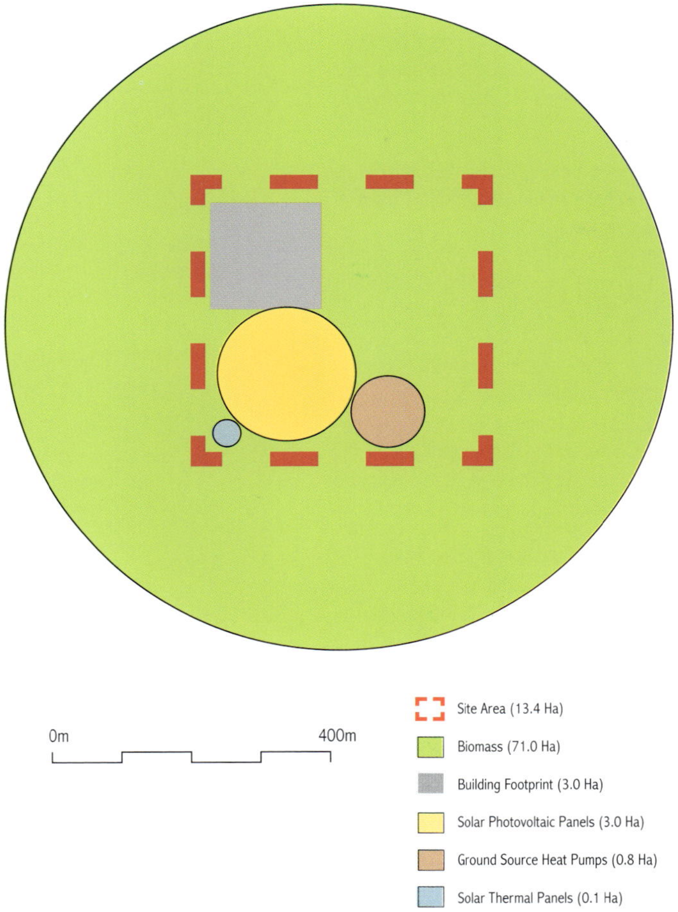

Figure A21 - Space requirements for energy technologies

Energy and Resources

Figures A22 and A23 show the amount of roof space required for solar PVs to achieve zero carbon and a similar application in Germany. Figure A24 also shows the area set aside for a network GSHP array.

Figure A24 shows a modern overshot waterwheel in operation in Germany. In Hackbridge a modern undershot wheel would be more suitable.

The Water Solution

To reduce the water consumption from 110 litres per person per day (the level of consumption set by the Code Level 3) to the target 80 litres per person per day set by the Code Level 6. A further saving of 30 litres per person per day would need to be provided from rainwater harvesting.

If the number of consecutive dry days is likely to be a maximum of say 20 the total required harvesting would be approximately 1250 m^3. If it is only five days then 310 m^3 storage must be provided.

On the basis of five days, nine 35 m^3 tanks would be required within the Felnex site to achieve sufficient capacity to meet Code Level 6.

Figure A25 shows a communal rainwater harvesting tank located within each courtyard development area to serve the needs of the surrounding homes. These would be between 30 m^3 and 40 m^3 and located within the rain gardens to receive water from the SUDS system. These tanks would supply over half the local water flushing and irrigation requirements, with surplus storage allowing capacity during flooding.*

Figure A22 - Solar Panels at Vauban, Germany

Figure A24 - Modern Waterwheel, Bavaria

* No storage amount is added to the total SUDS calculation on the basis that the rainwater harvesting tank could be full at the time of the storm.

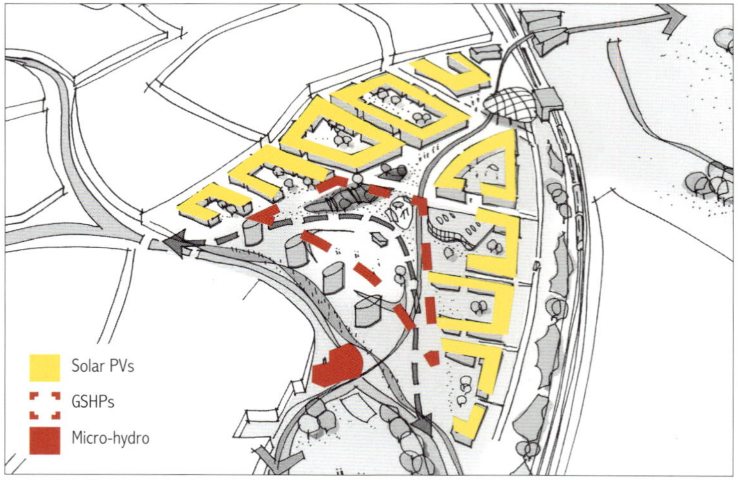

Figure A23 - Locations of renewable energy infrastructure

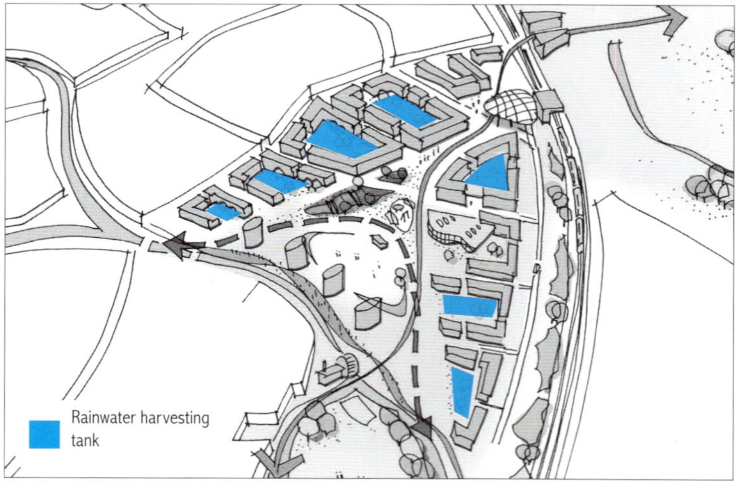

Figure A25 - Rainwater harvesting areas

The flashy nature of the river flows but low volumes of water mean that diverting water through the Felnex site could help to alleviate flood flows downstream, by attenuating the flood hydrograph. However, it is important that this does not slow flood flows onto site thus causing flooding upstream.

The flood risk management solution for Hackbridge is to slow and store a portion of the floodwater as it passes the Felnex site to reduce the peak flows, ie slowing the flow of water passing through the site to decrease the height of the floodwater and increase the warning time, particularly to sites downstream.

This would be achieved by diverting water from the river, as it nears bank full, into a flood storage area on the Felnex site via a backwater channel. The flood storage area is located in a natural low point on the site that is enlarged and embanked to create a deeper pond that can be used for storage.

In addition the riverbank would be widened and lowered to create a more natural profile and additional flood capacity. This has the benefit of reducing the flooding that might occur from more frequent events.

Figure A26 shows the flood zones for the site.* The area closest to the railway station is Flood Zone 1, therefore suitable for all development uses in accordance with Tables D.2 and D.3 of PPS25. The remainder of the site is mostly Flood Zone 2, therefore it is suitable for residential development. A small area is shown in Flood Zone 3. Development in this area would require floodplain compensation.

The proposed development is predominantly residential and therefore suitable on the site.

Climate change could result in greater levels of inundation. Therefore the EA required floor levels to be set 400 mm above the current (1 in 100 year) flood level plus freeboard. The threshold swales and roads allowed level differences to be created to achieve these levels.

The area of Flood Zone 2, which is closest to the river, is most sensitive to change in future flood levels. Therefore this would be used for offices, not residential.

The potential sources of flooding are indicated in figure A27.

* Since this work was carried out the SFRA level 1[31] was published with updates showing a more extensive flood zone 3, similar to that indicate in figure A10.

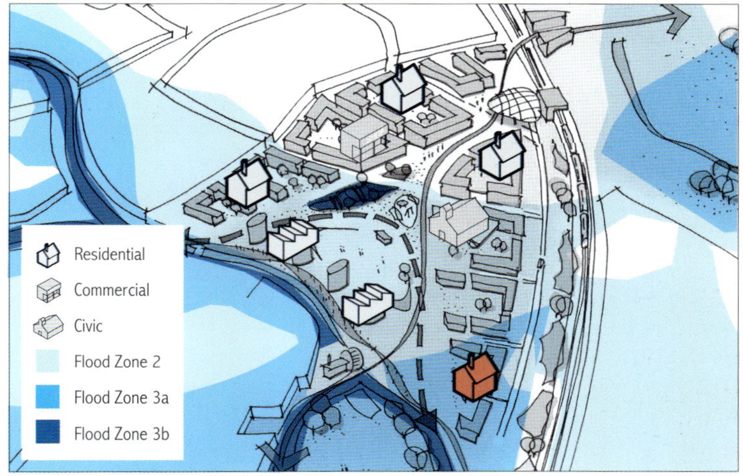

Figure A26 - Development and flood risk

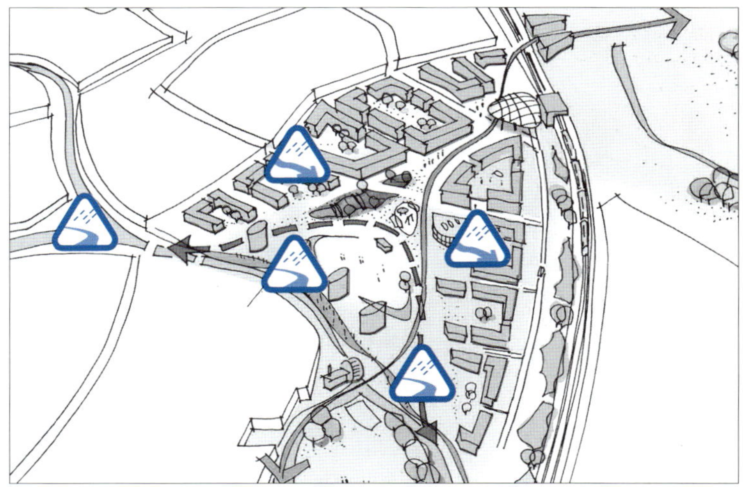
Figure A27 - Potential sources of flooding

Flood Risk Management - SUDS

Because the two main sites at Hackbridge were previously developed (ie brownfield) the surface water drainage requirements are likely to be to mimic prior surface water flows and not to provide for infiltration and attenuation to replicate 'greenfield' levels.

The EA policy aim is to replicate 'greenfield' surface water drainage rates where possible.

The local geology, depth to groundwater and contamination affect the feasibility of an infiltration system. The likely predominance of clays could result in an infiltration system being largely ineffective and the likelihood of contamination from previous industrial uses would make infiltration unsuitable. Therefore the SUDS solution is to install source control measures (such as the swales shown in figure A28) with storage areas outside Flood Zone 3 to attenuate the run off, reducing the risk of flooding on site and slowing the water flowing into the river.

Figure A29 indicates a number of 3 m wide swales and generous rain gardens that would allow large volumes of rainwater to be held on site (5000 m^3). These areas would also be used for amenity and provide opportunity for environmental gain.

The SUDS system would be linked to the main flood drainage paths to discharge water after a heavy downpour into the watercourse (figure A30).

Large communal rainwater harvesting tanks would provide capacity to reduce water consumption and provide additional surface water storage. These would be located beneath play areas located at the centre of the courtyards and rain gardens.

Figure A28 - Gravel swale

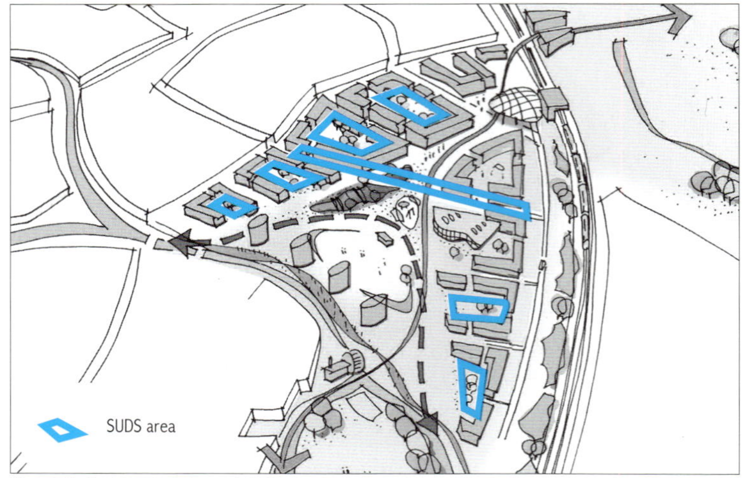

Fiure A29 - Sustainable drainage detention basins, rain gardens and swales

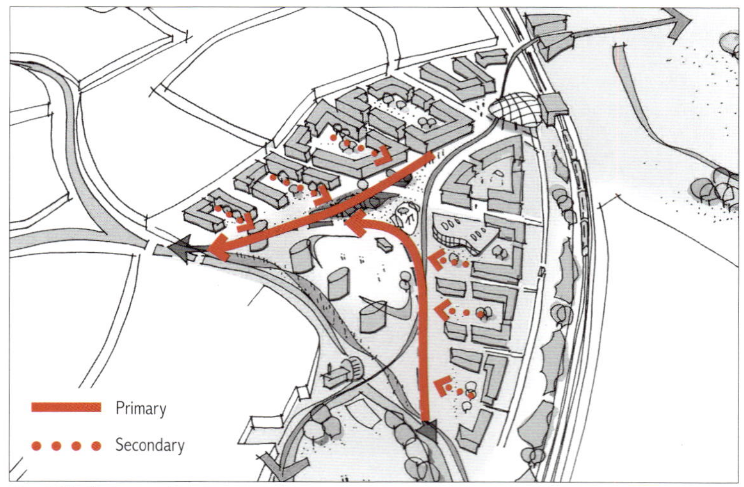

Fiure A30 - Drainage paths

Safe Design

Figure A31 illustrates the location of the fluvial flood areas and shows the overlay with the exit routes.

Pathways through the village blue/green are raised above the fluvial stores, allowing safe egress from the park during a flood. These paths connect with the higher roads that direct residents away from the site to higher ground.

At the centre of the development the school building offers refuge to residents.

Strategic Flood Alleviation Measures (outside study area)

Diversion to the east of the railway from upstream of the development site to link to ponds associated with Beddington sewage treatment works appears in plan to be a feasible option to divert flood flows away from Hackbridge. The cost of such a scheme would most likely be high and the logistics to confirm a route would be difficult. However, this could provide a long-term option to reduce flood risk in the areas.

The most direct path, avoiding major infrastructure, is shown in figure A32. This would currently run through existing residents' back gardens. Therefore it is unlikely that this type of measure could be delivered without the support of the local residents and authority. An alternative route may be possible.

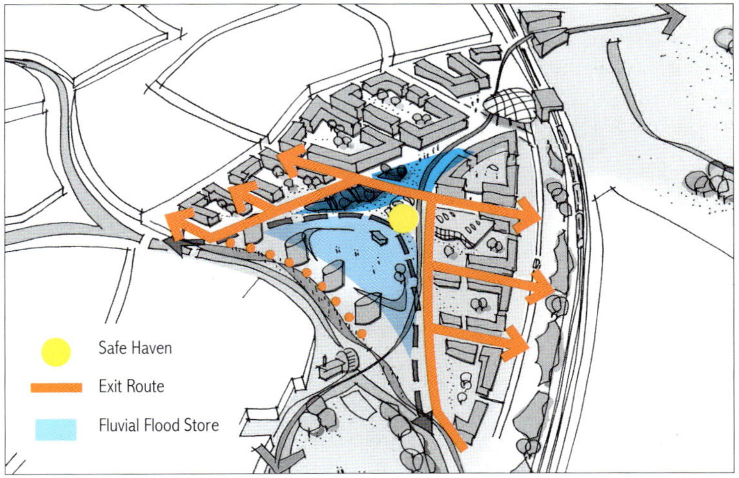

Figure A31 - Safe havens, evacuation routes and flood stores

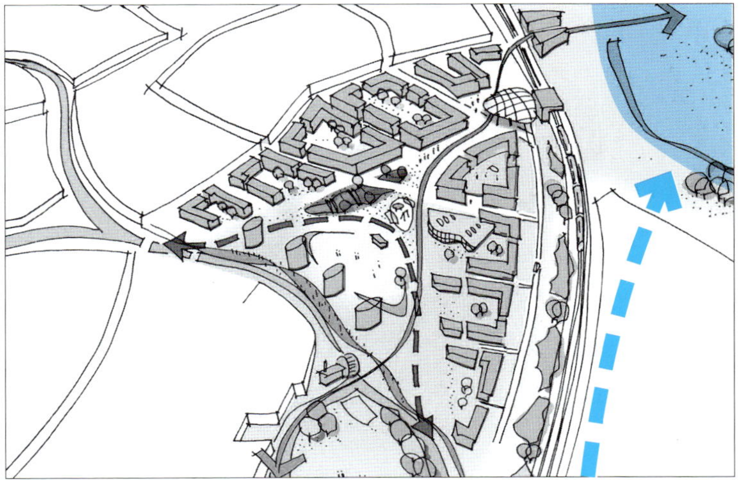

Figure A32 - Strategic flood measures

Landscape

It was identified that the area of open space and public amenity required for the development proposed was greater than the total area of the site.

To achieve a high quality suburban environment land was planned to be multi-use, with different planting characteristics and mowing regimes providing apparent subdivision. In particular the central village blue/green was planned to provide areas for formal sports, informal play, public parkland and educational stimulus in the form of the dry ponds and planted swales, shown in figure A34.

High-quality native trees and herbaceous planting would be provided in rain gardens, market squares and streetscape around development.

Reed beds and marginal vegetation would be provided in the wetland area and sections of the river corridor; with aquatic vegetation in the river and areas of standing water in the wetland.

Figure A33 illustrates the areas within the development for outdoor play.

Each courtyard would provide a local area for play (LAP), shown in figure A35.

The village centre would include a neighbourhood equipped area for play (NEAP) and multi-use gaming area (MUGA)

The village green would include a dedicated local equipped area for play (LEAP), adjacent to the school.

Small allotment plots would be provided adjacent to the school to encourage learning to grow food.

Figures A36a to A36d show examples of other sites where similar ideas have been implemented.

Figure A37 shows an indicative development plan. Figures A38a to A38c show a section through this plan.

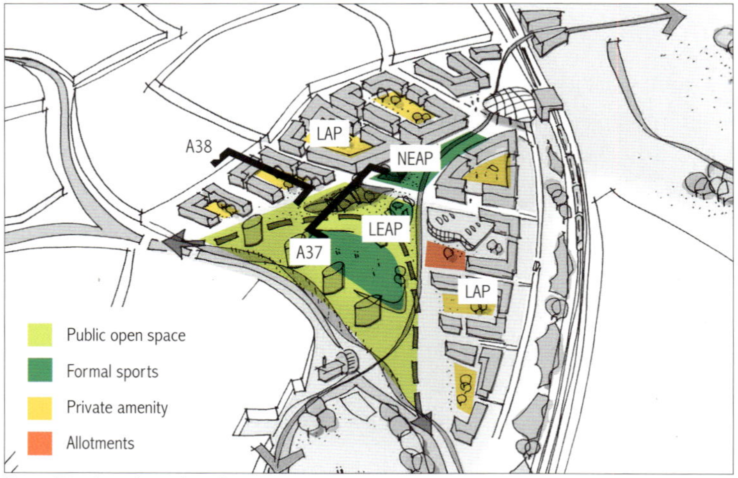

Figure A33 - Areas for outdoor play

Figure A34 - Dry pond and swales

Figure A35 - Local areas for play combined with rain gardens and rainwater harvesting

Figure A36a - Landscaped flood alleviation channel, Mill Creek Canyon, USA

Figure A36b - Swale in Malmo, Sweden

Figure A36c - Ecological pond

Figure A36d - Preston Bagot, shallow lake

Indicative Site Plan

'Village Blue'
- Dry pond and some wetland habitats for biodiversity gain
- Accessible via boardwalk for amenity and educational value
- Soft character, using native tree and plant species

Rain Gardens
- Shared/communal amenity space
- Storing/treating rainwater and surface run off from housing

Outflow to river

'Causeway'
- allowing flood water to spill into multi-functional green space for attenuation

Room for the River
- Extension of the Wandle Trail
- Amenity landscape corridor focused on shared footpath/cycleway
- Riparian and wet woodland habitats

Water Mills
- Electricity generation
- Visitor Centre

'Village Centre'
- Shared surface for vehicles, pedestrians and cyclists
- Multi-functional space for parking, markets, assembly and one-off events
- Hard surface sports area with capacity for attenuation

Paved 'event space' over control device for swale outfall

'Locally Grown Food'
- Allotments, or
- Community gardens, or
- Community orchard

'Village Green'
- Amenity open space
- Informal recreation
- Potential for organised sports
- Capacity for attenuation

Floodwater from river conveyed to wetland via swale

Figure A37 - Site plan

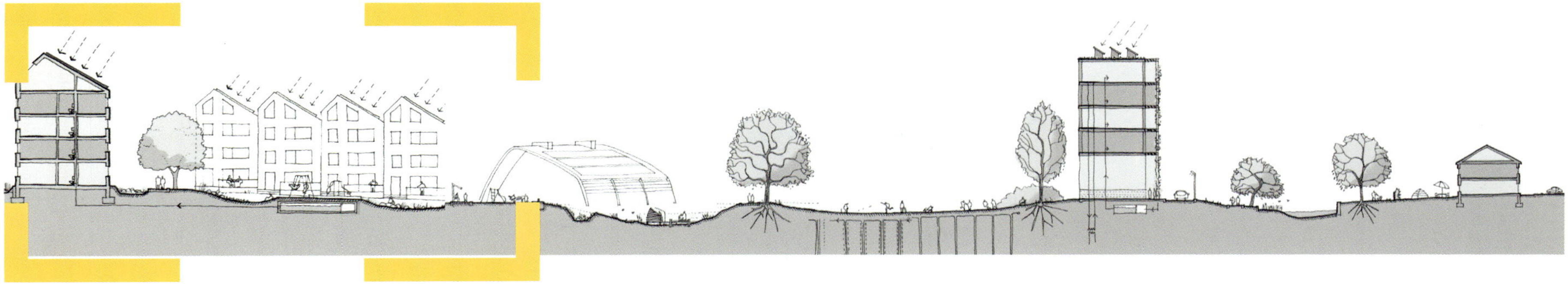

Figure A38a - Section through a courtyard rain garden

Indicative Section

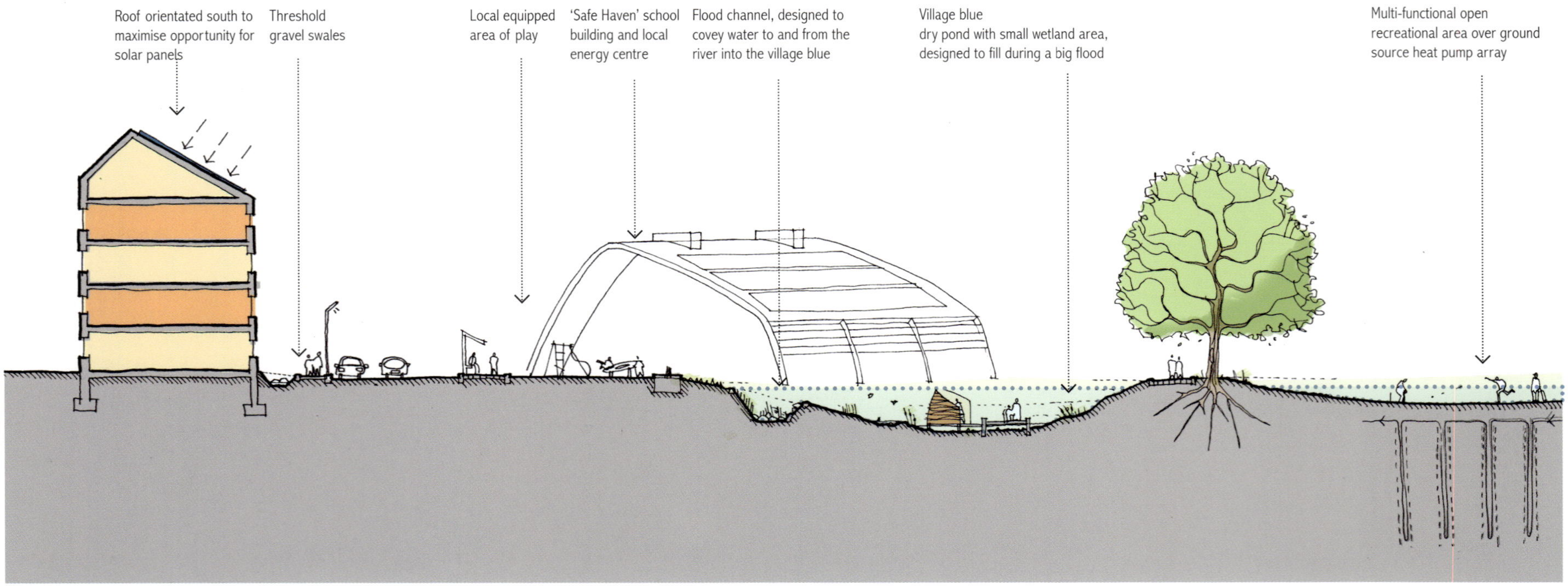

Figure A38b - Section through the Village Blue/Green

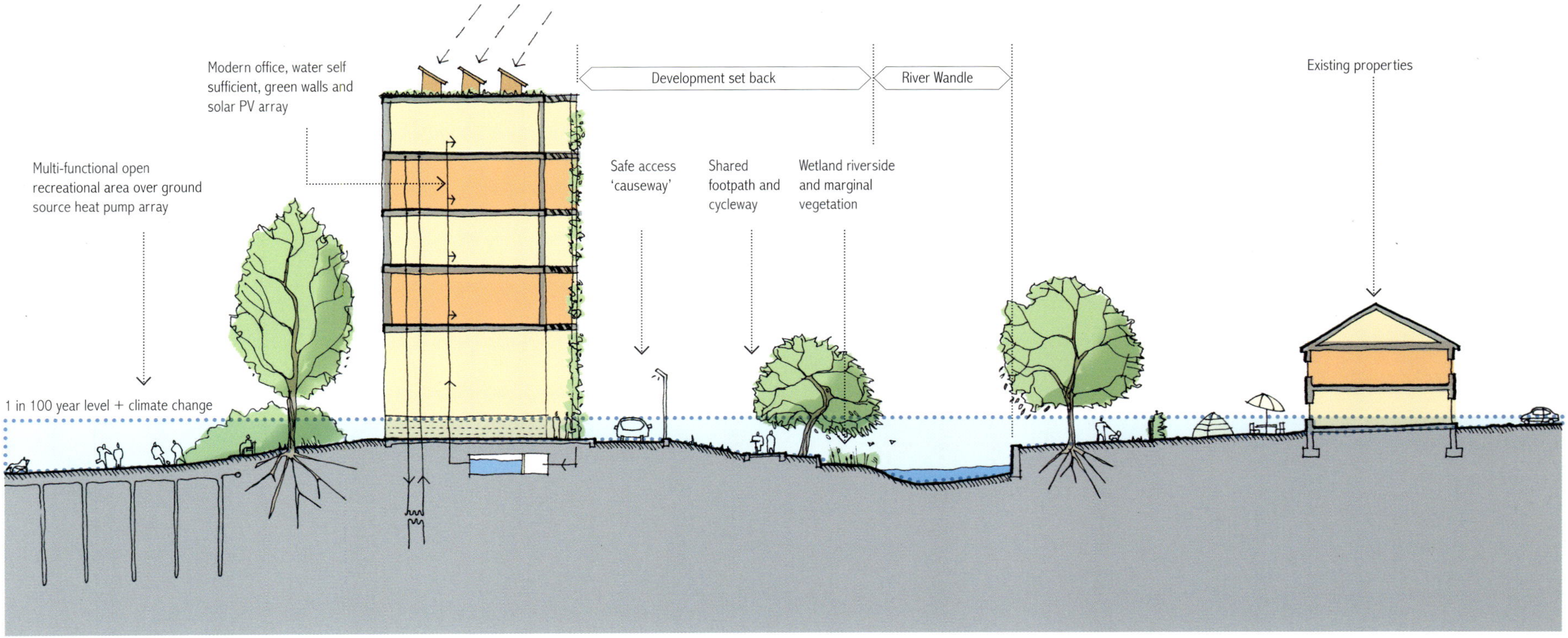

Figure A38c - Section through the River Wandle

Figure A39b - During a 1 in 20 year storm the rain gardens and swales retain rainwater, reducing run off overland and into the sewers.

Figure A39c - During a 1 in 100 year storm rain gardens attenuate more rainwater and floodwater is directed onto the village blue/green, away from people's homes.

Figure A39a (left) - View from the rain garden courtyard, looking towards the distant river, and village blue/green.

Review >

Costs

The design proposals for all three case studies are assessed collectively in the main report (pages 26-29). A summary of the main issues, particularly with regard to flood risk management, is given below.

Affordability

The total cost of the development for the Felnex site, built to zero carbon, Code for Sustainable Homes Level 6 was calculated at approximately £176,000,000. This was projected as an increase of 34% over a traditional development built to current regulations (2008).

Table A5 and figure A40 show a breakdown of the costs.

The cost uplift was primarily a result of the renewable energy infrastructure costs, which were assessed to be four times the cost of traditional energy infrastructure. The second largest increase in cost was £7.5 m associated with other measures designed to raise the standard of construction from Code for Level 3 to Level 6. These include items such as building materials, building, rainwater recycling and site management.

Site 1, Development Costs

	Base Option Costs	LifE Option	Percentage change from base (LifE Option)
Construction	£94,136,885	£101,339,965	7.65%
Transport and Land Use	£21,181,200	£19,427,590	-8.28%
Flood Management	£1,288,300	£1,250,040	-2.97%
Energy and Utilities	£10,961,450	£42,223,160	285.20%
Other Code Measures	£3,737,700	£11,690,540	212.77%
TOTALS	£131,306,000	£175,932,000	33.99%

Cost plan exclusions:

1) Professional fees
2) Land acquisition costs
3) S106 & S278 costs
4) Service diversions
5) Protection of existing structures
6) VAT
7) Inflation beyond 1Q 2008
8) Costs arising from archaeological investigations
9) Demolition of existing structures on the site
10) Public consultation costs
11) Legal fees
12) Finance charges
13) Sales and marketing costs
14) Compensation costs
15) Amounts payable for continuing maintenance & commuted sums
16) Life cycle & maintenance costs including to beach and river defences
17) Land remediation costs

Table A5 - Cost summary

Costs

Fluvial Flood Risk Management

The cost of excavating areas of the site to provide additional flood storage and drainage channels was found to be approximately £500,000 less than the cost to elevate properties out of flood risk and excavate areas off site to compensate for loss of floodplain. However, to add resilient building measures to the design increased the cost by just under £6.5m. Therefore the total cost would be £6m more than raising land and compensating for the loss of floodplain.

The difficulty with this approach is that there were no obvious sites available to provide 2.2 hectares (5.5 acres) of compensatory excavation and if there were the price of this land could be artificially elevated to prevent the development occurring.

Integrated Approach

The cost of developing the site to Code Level 6 and combining ecological flood mitigation in a non-integrated way was assessed to be slightly more expensive than with the LifE integrated approach, see figure A41. The integrated approach was found to save up to £1.75m, 1.0% of the total development cost.

This assessment was made on the basis that without establishing an integrated plan the area of land required for parking, recreation and SUDS would be greater unless land uses were combined and development could be located close to public transport. Therefore multi-use amenity space was assessed to provide a financial benefit.

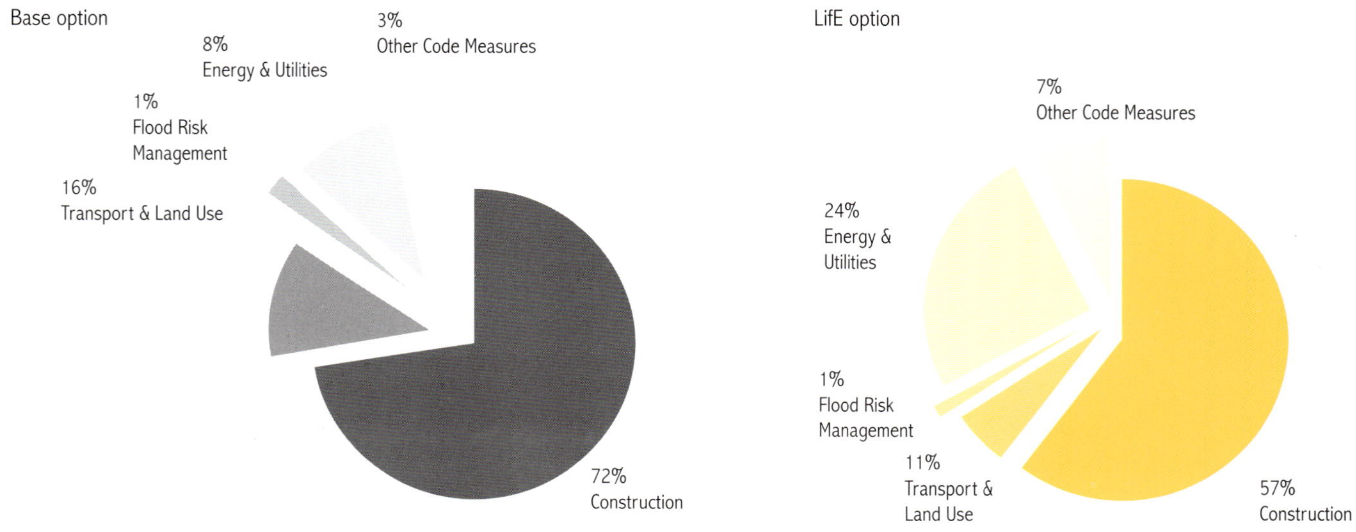

Figure A40 - Cost comparison between base and LifE option

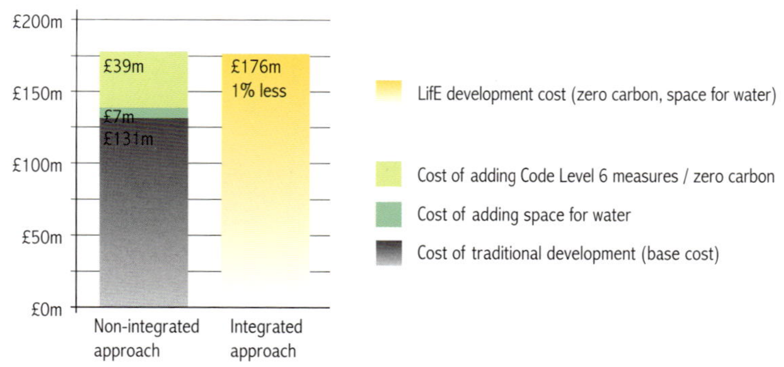

Figure A41 - Cost between integrated and non-integrated approach

Acceptability

The LifE project proposals were assessed against related planning policy, relevant building regulations, the Code for Sustainable Homes and a number of other guidance documents.

In general the proposals were found to be in accordance with policy. The positive integration of environmental improvements and flood risk management with zero carbon development was well received by the local authority; and complemented their aspiration for a sustainable suburb.

An apparent failing of the development proposals was the lack of amenity provision. This was due to the density of development proposed (considered medium development) on the site and the difficulty of meeting play area standards first established in 1925. In these situations it is common for the developer to be granted planning permission on the basis of providing a financial contribution towards improvements to existing park facilities. Therefore the amount of provision suggested within the development was thought to be acceptable.

The Sequential Test

Approximately 50% of the total site area (of all three sites) lay within Flood Zone 1 and 40% within Flood Zone 2. Therefore the majority of the site was found to be suitable for development. There were no obvious alternative sites available in the area therefore the areas within Flood Zones 2 and 3 could also be considered.

Redevelopment of the Wandle riverside site would be appropriate for commercial use, such as offices, in accordance with PPS25 Table D.2. However, if it were to be considered for residential development then the exception test would need to be passed.

A high-level flood risk assessment is shown in Table A6 (page 82).

Appropriateness

The industrial units within Hackbridge were nearing the end of their useful lives and many start-up businesses were relocating to larger premises. The shortage of affordable and 'first-time buyer' homes in the area and lack of available sites meant that these former industrial sites were under pressure for redevelopment. The two sites closest to the railway station (Felnex and Bedzed North) were well suited for sustainable residential development with reduced car dependency. The Wandle riverside site was less suitable, due to the increased travel distances but was still suitable for commercial or mixed-use development. It was thought important to retain commercial uses within the area to provide jobs and to contribute to the economic sustainability. Therefore it was considered appropriate to provide mixed-use development of this site.

Adaptability

The greatest effect of climate change on flood risk within site 1 would be from increased intensity of rainfall. The current urbanisation of the river catchment and the increase in development located near to the river has resulted in an increased vulnerability to flooding. This vulnerability could increase with more intense rainstorms.

The improvements to the surface water management and river edge suggested on the Hackbridge sites could help to reduce the risk on site. Further urbanisation of the catchment could diminish this benefit and result in the flood risk rising. Therefore the flood resilient building measures proposed as future proofing (such as sockets at higher levels and horizontal plasterboard), were thought sensible.

Flood Risk Assessment

1	**Development description and location**	**The River Wandle at Hackbridge**
1a	What type of development is proposed and where will it be located?	Mixed-use development for a range of sites extending from the river to the railway station
1b	What is its vulnerability classification?	Combination of more and less vulnerable
1c	Is the proposed development consistent with the Local Development Documents?	Yes
1d	Please provide evidence that the Sequential Test and where necessary the Exception Test have been applied in the selection of this site for this development type?	N/A. The three study sites were selected during Stage 1 of the project from a number of real sites submitted by local government from around the country. The three sites were chosen on the basis of providing a range of detailed issues to test the 'LifE' strategy and providing a number of flood risk issues that would be most representative of issues likely on other sites. Being a redevelopment/regeneration site, improving the existing flood risk situation and providing wider sustainability benefits means the Hackbridge site would likely pass the Exception Test.
2	**Definition of the flood hazard**	
2a	What sources of flooding could affect the site? (see annex C PPS25).	Flooding from the River Wandle Flooding from land Possible flooding from sewers
2b	For each identified source, describe how flooding would occur, with reference to any historic records wherever these are available.	Typically following heavy rain with fluvial flooding of short duration with rapid rise and fall of river levels; surface water flooding most often occurs following high intensity summer rainstorms when the drainage system cannot cope.
2c	What are the existing surface water drainage arrangements for the site?	Unknown. Not investigated as part of this study. The scope of work for this study did not involve the development of full FRAs for each of the three sites.
3	**Probability**	
3a	Which flood zone is the site within?	Flood Zones 1, 2 & 3
3b	If there is a Strategic Flood Risk Assessment covering this site, what does it show?	Yes, but only Stage 1 assessment to date.
3c	What is the probability of the site flooding, taking account of the contents of the SFRA and of any further site-specific assessment?	Predominantly medium probability – the probability of flooding is unlikely to change as a result of the SFRA.
3d	What are the existing rates and volumes of run-off generated by the site?	Unknown. Not investigated as part of this study. The scope of work for this study did not involve the development of full FRAs for each of the three sites. The development sites, however, are previously developed with a high percentage of hard surfacing.
4	**Climate change**	
4a	How is flood risk at the site likely to be affected by climate change?	Increased flows will increase the depth, extent and volume of the river flooding for a particular frequency event. However, the future extent of the 100 year event has not been modelled. The EA has recommended using the 1000 year event to indicate the extent of the 100 year future event.
5	**Detailed development proposals**	
5a	Please provide details of the development layout, referring to the relevant drawings.	See drawings
5b	Where appropriate, demonstrate how land uses most sensitive to flood damage have been placed in areas within the site that are at least risk of flooding.	Higher density of residential development has been located close to the railway station and in Flood Zone 1. A proportion of residential is located in the current Flood Zone 2.
6	**Flood risk management measures**	
6a	How will the site be protected from flooding, including the potential impacts of climate change, over the development's lifetime?	Flood defences are not proposed but compensatory floodplain storage will be incorporated into the development, and there are opportunities for river channel restoration. Site layout and compensatory storage volumes take account of climate change.
7	**Off-site impacts**	
7a	How will you ensure that your proposed development and the measures to protect your site from flooding will not increase flood risk elsewhere?	See above, no defences and floodplain storage will ensure no increase in flood flows upstream or downstream. SUDS systems will provide for 'no detriment', and in all probability improvement, to the surface water flooding situation.
7b	How will you prevent run-off from the completed development causing an impact elsewhere?	See 7a above.
8	**Residual risks**	
8a	What flood-related risks will remain after you have implemented the measures to protect the site from flooding?	See above, the site will not be protected from flooding. Flood risk will increase with climate change and the development planning, location, level and design of buildings, takes this into account.
8b	How, and by whom, will these risks be managed over the lifetime of the development?	Environment Agency = river channel maintenance; Floodplain storage ponds, SUDS systems, with local community/authority management

Table A6 - Flood Risk Assessment (FRA)

Conclusions

The proposals for the sites at Hackbridge showed that the flood risk could be managed with non-defensive measures to reduce the level of risk to the sites and potentially surrounding area. These measures included:

- avoiding risk by locating the highest density development in the areas of lowest probability of flooding and locating less vulnerable uses in the higher risk areas;
- reducing the extent of flooding by reinstating areas of the floodplain, providing areas of flood storage and rainwater attenuation;
- mitigating the effects of flooding by use of resilient building construction and providing dedicated channels to return floodwater to the river channel after a flood.

The areas of flood storage and rainwater attenuation provided additional opportunity for environmental gain, recreation and low carbon energy generation in the form of ground source heat pumps.

It was not found possible to create a zero carbon development by solely using on-site renewable technologies. All available roof space was found to be necessary for solar panels, precluding the opportunity to use green roofs for rainwater attenuation. The lack of available energy sources, combined with the lack of space for photovoltaic solar panels resulted in the need to use off-site renewable energy.

Safe redevelopment of the Felnex site was considered feasible by setting development back from the river, providing access from the adjacent roads and resilient construction closest to the river. Floodplain restoration and rainwater attenuation was thought to be the most effective means of reducing flood risk to the site. Rain gardens and swales located within development courtyards provided sufficient space to attenuate rainwater for a 1 in 100 year storm.

Safe redevelopment of the Riverside site was considered feasible by aligning buildings with the river, creating raised access and using resilient construction. Rainwater attenuation above ground, through use of green roofs, was found to potentially conflict with the need to use roofs for solar panels. Provision of green roofs was therefore only considered within Flood Zone 3.

There was insufficient space on site to provide the outdoor play requirements for the expected number of residents. Therefore multi-use of the land available was essential to providing a reasonable provision of recreation space.

Integration of flood mitigation measures; reduction in car parking provision and use of the floodplain enabled higher numbers of zero carbon homes to be achieved than would be otherwise possible.

Figure A42 illustrates the spatial requirements for the Felnex site and the various provisions for amenity space and flood mitigation.

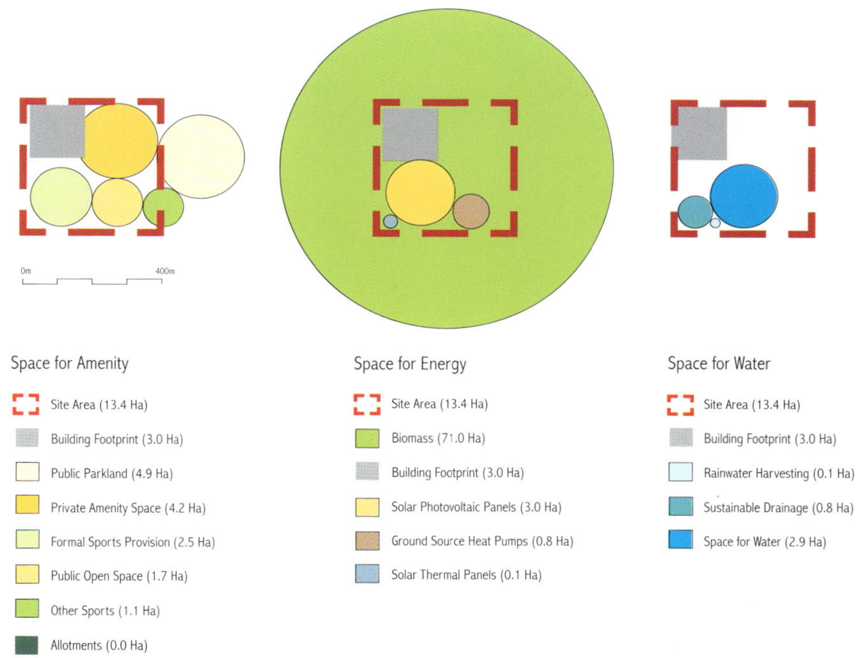

Figure A42 - Space demands for amenity, energy and water on the Felnex and rail site in Hackbridge

The upper catchment

The key message for this site was to slow the flow of rainwater during peak rainfall, allowing time for peak river flows to pass before discharging into the water course.

'let rain slow'

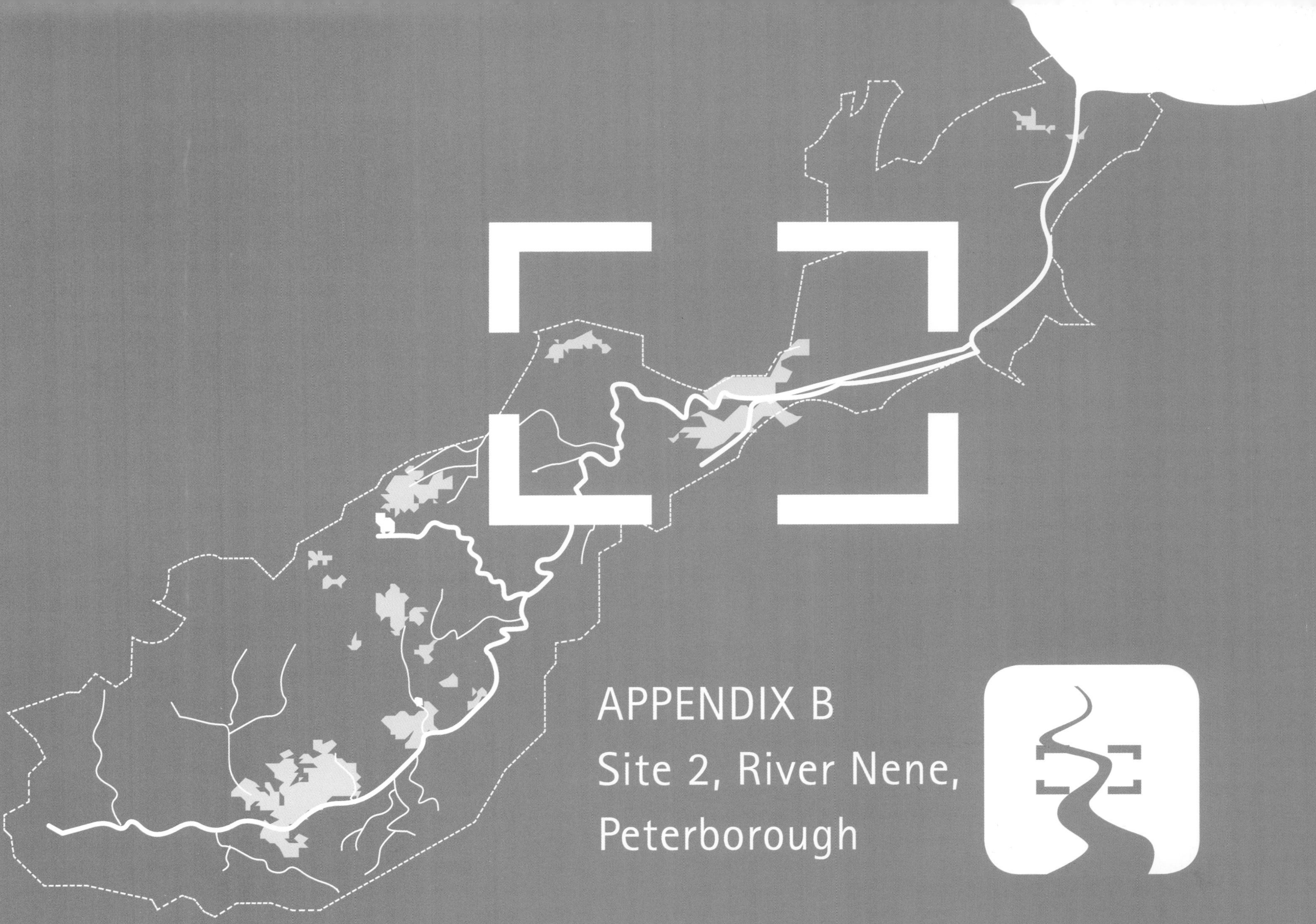

APPENDIX B
Site 2, River Nene, Peterborough

Contents

SITE 2 > The River Nene at Peterborough

The River Nene at Peterborough

Site Analysis >

SITE 2 > The River Nene at Peterborough

Site 2
Location....................East of England
River...........................Nene
Catchment.................Middle Catchment
Current use...............Brownfield
Area...........................50 hectares
Units planned............2400
Density......................48 units/hectare

Understanding the Location

Peterborough is located in the East of England, at the northern edge of the Fens and situated on the River Nene (figure B1). Peterborough is a main line station on the East Coast Main Line railway connecting London and Edinburgh. Direct services to London take between 45 and 60 minutes and run every 20 minutes during peak time.

Peterborough City and its surroundings have been identified by the former ODPM for major development of homes and business as part of the London-Stansted-Cambridge-Peterborough growth area.[41] Approximately 25,000 homes are planned by 2016 of which the Riverside sites could potentially provide 2000 of these.

Opportunity Peterborough, an urban regeneration company (URC) has been tasked with leading the regeneration of Peterborough. Their aim is for the City to become one of England's greenest.

The Site, referred to as Peterborough Riverside, is a collection of sites, covering 77 hectares (of which 50 hectares were deemed developable), on the banks of the River Nene and close to the Cathedral and historic city centre. The site includes brownfield, former landfill and areas currently used for civic leisure and recreation, including a theatre and lido set within 'Central Park'.* An English Partnerships Carbon Challenge site is located immediately to the south.

The majority of the site is on the north bank, bisected by the A1139 elevated roadway, which runs north-south over the river. Existing infrastructure boundaries have been used to divide the site into sub-areas, as shown in Figure B2.

- West Rivergate is currently occupied with a mix of retail and residential buildings
- West Embankment is the city park and sports grounds
- South West Embankment contains the City electricity sub-station.
- North South Embankment is used for light industry and offices and is part vacant
- The Fengate site was formerly used by the sewage treatment works and for landfill.

Figures B3a to B3f show photographs of the site and surrounding area.

* Not all of the sites were (at the time of the study) being brought forward for development. However, they were included within the study to allow consideration of a wider strategic plan.

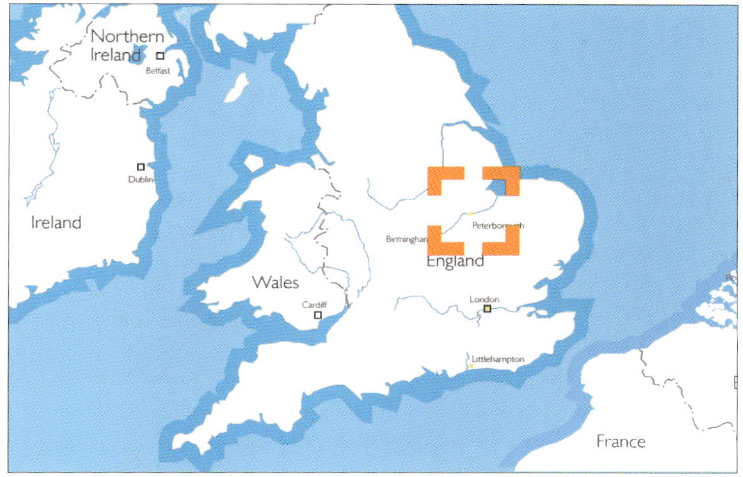

Figure B1 – Location plan

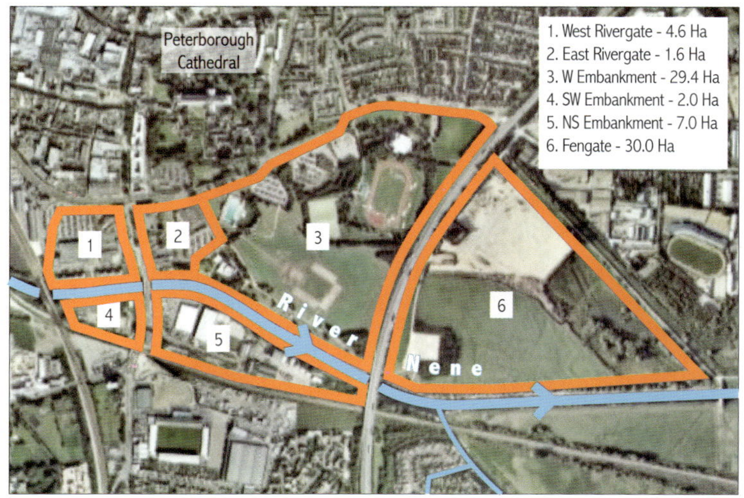

Figure B2 – The site areas

Site Photographs

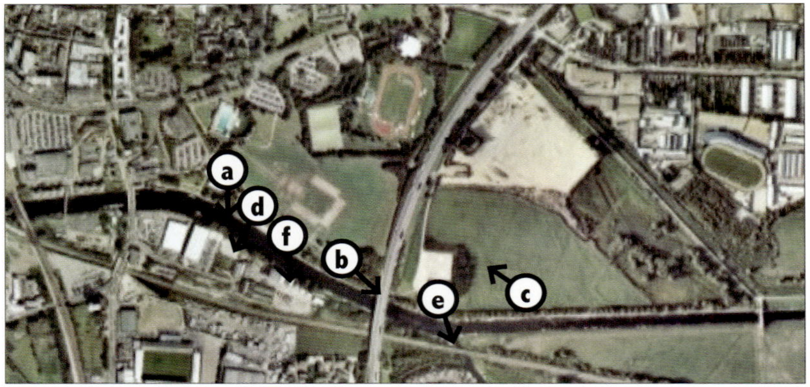

Figure B3 - Photo location plan

Figure B3c - View from the Fengate site towards the elevated roadway

Figure B3a - View from Public Park

Figure B3b - View of elevated road and bridge

Figure B3d - The listed warehouse on the south bank

Figure B3e - Connection with 'Back River Drain'

Figure B3f - View of south bank of the river

Peterborough is an important historic settlement dating back to the Bronze Age. A monastery was founded in 655 and Peterborough Cathedral was completed in 1238.[42]

Peterborough was an important area of ceramic production in the Roman period, providing Nene Valley Ware that was traded as far away as Cornwall. The population grew rapidly following the arrival of the railways in the nineteenth century, and Peterborough became an industrial centre, particularly noted for its brick manufacture.[42]

The 1898 Ordnance Survey map shown in figure B4 indicates that the Great Eastern Railway Station was located immediately to the south of the river within the 'North South Embankment site'. Through the centre of what is now central park a small water channel ran west to east to a sewage pumping station. The land to the east is indicated as 'liable to floods'.

The 1936-52 Ordnance Survey map shown in figure B5 indicates that a recreational ground and Lido were established and a hospital for infectious diseases.

A number of buildings and parks within the area have received statutory listing. These are highlighted in figure B5 and listed below:

Scheduled Ancient Monument

1. Peterborough Cathedral Precincts, including Table Hall and Infirmary Arcade (PE140)

2. Old Customs House (PE195)

Statutory Listed Building

3. The Lido, Bishops Road (Grade II)

4. Old Customs House, Embankment Road (Grade II)

Numerous listed buildings in and around Peterborough Cathedral Precincts

Register of Parks and Gardens of Special Historic Interest

5. Peterborough Minster Precincts (Grade II)

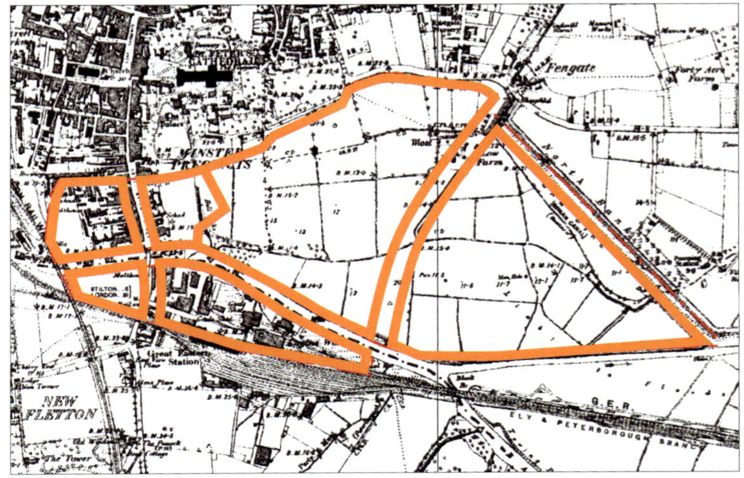

Figure B4 - 1898 Ordnance Survey map

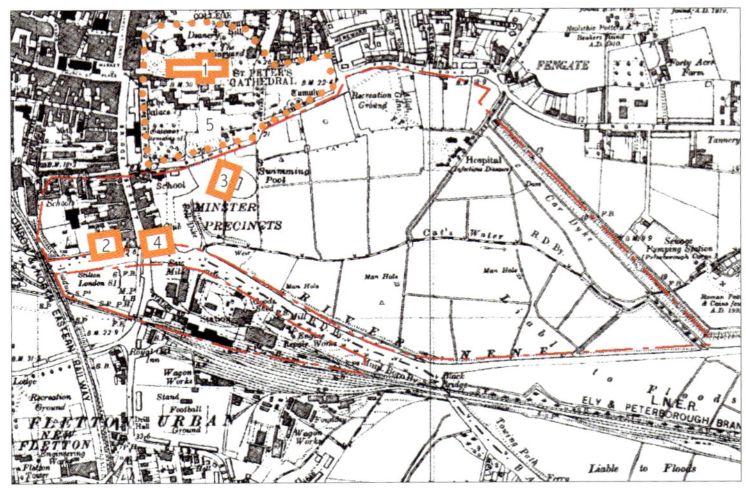

Figure B5 - 1936 - 52 Ordnance Survey map

The River Nene

Peterborough is at the lower end of the middle catchment of the River Nene. The Nene is the ninth longest river in the United Kingdom and navigable from Northampton to its mouth, linking the River Great Ouse with the Grand Union Canal.

The Nene flows from its source in Arbury Hill (in Northamptonshire) through a number of significant towns (including Northampton, Kettering, Wellingborough, Corby and Peterborough) and out to the Wash and North Sea.

The River Nene is a well-managed river system with large areas of designated floodplain within the middle and lower catchments. The lower catchment has been significantly altered to improve flood management and river navigation. The river now bifurcates within the boundary of the site and the river flows down an artificial channel towards the 'Dog-in-a-Doublet' tidal sluice some 7 km downstream. To the south of the artificial channel a 20 km-long strip of land, known as the Nene Washes has been reserved to provide fluvial flood alleviation and fluvial flood storage during high tides, before discharging river water back into the main channel during low tide. Until the construction of the Dog-in-a-Doublet sluice in 1936, the River Nene was tidal at least as far upstream as Peterborough.[35]

Constant rainfall in spring 1998 resulted in the river overflowing its banks and flooding many parts of Northampton and Wellingborough. Flooding in Peterborough was limited to parts of Central Park, including the theatre, some sections of the riverside to the north and the low lying land to the east of the A1139, within the Fengate site.

The topographic map in figure B6 shows the river channel running through the flat land to the south of Peterborough city centre.

Figure B7 shows the extent of historic floods.

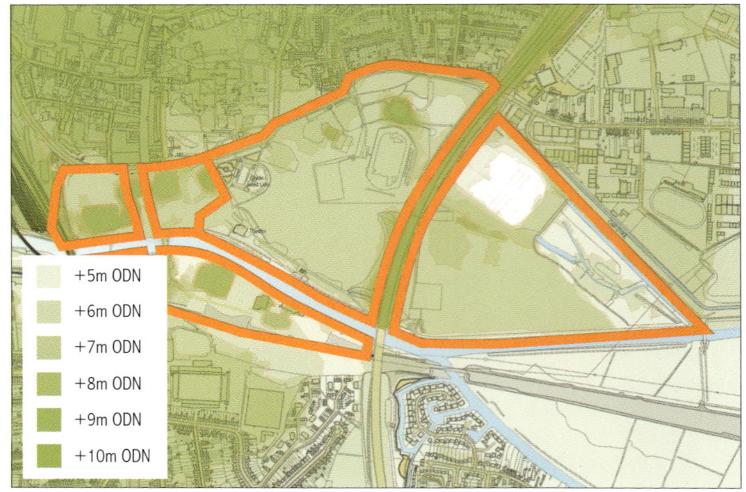

Figure B6 - Topographic map

Figure B7 - Historic flood extents

The Nene Valley

The character of the Nene Valley changes greatly as it moves from the shallow undulations of the clay/limestone junction west of Peterborough to the flatter fen edge to the east (see Figure B8).

To the west of Peterborough lies the Ferry Meadows Country Park and associated recreational facilities. The intensive organised landscape around the lakes gradually changes to a pastoral agricultural use that still retains extensive areas of flood meadows and associated sites of nature conservation interest.

East of Peterborough the landform flattens and the river has been adapted to the rectilinear drainage pattern of the fens.

The old course of the River Nene meanders south to disappear in a series of small drains around Farcet; the new river channel carries the river directly east across the fens through large open arable fields with sparse tree cover.

The Nene Washes

The Nene Washes (shown in Figure B9), located immediately downstream of Peterborough, cover an area of 1517.49 ha. They were established to provide a flood storage area for the Nene river flows when the tide level is greater than the level of the river, preventing water from flowing out to sea.

Cyclic use of the washes in the winter to store floodwaters from the river and in the summer for traditional cattle grazing has given rise to a mosaic of rough grassland and wet pasture, with a diverse ditch flora. In the winter the washes hold large numbers of waders and wildfowl, including the Bewick's Swan; and in the summer breeding waders and Spotted Crane. The washes act as a refuge for wildfowl and water birds displaced by deep flooding of the nearby Ouse Washes SPA and during severe winter weather elsewhere.

Car Dyke

To the east of the Fengate site a section of 'The Car Dyke' is evident (figure B9). The Car Dyke was originally thought to be a Roman canal, but is now believed to be part of a 1st or 2nd century AD fen drainage system. It follows a sinuous course down the south side of the Witham valley, surviving as earthworks for long stretches, but where levelled is also visible as cropmarks. The central ditch is banked on both sides, possibly forming part of a flood defence.[43]

Figure B8 – Nene Washes

Figure B9 - The Car Dyke

Landscape Characteristics

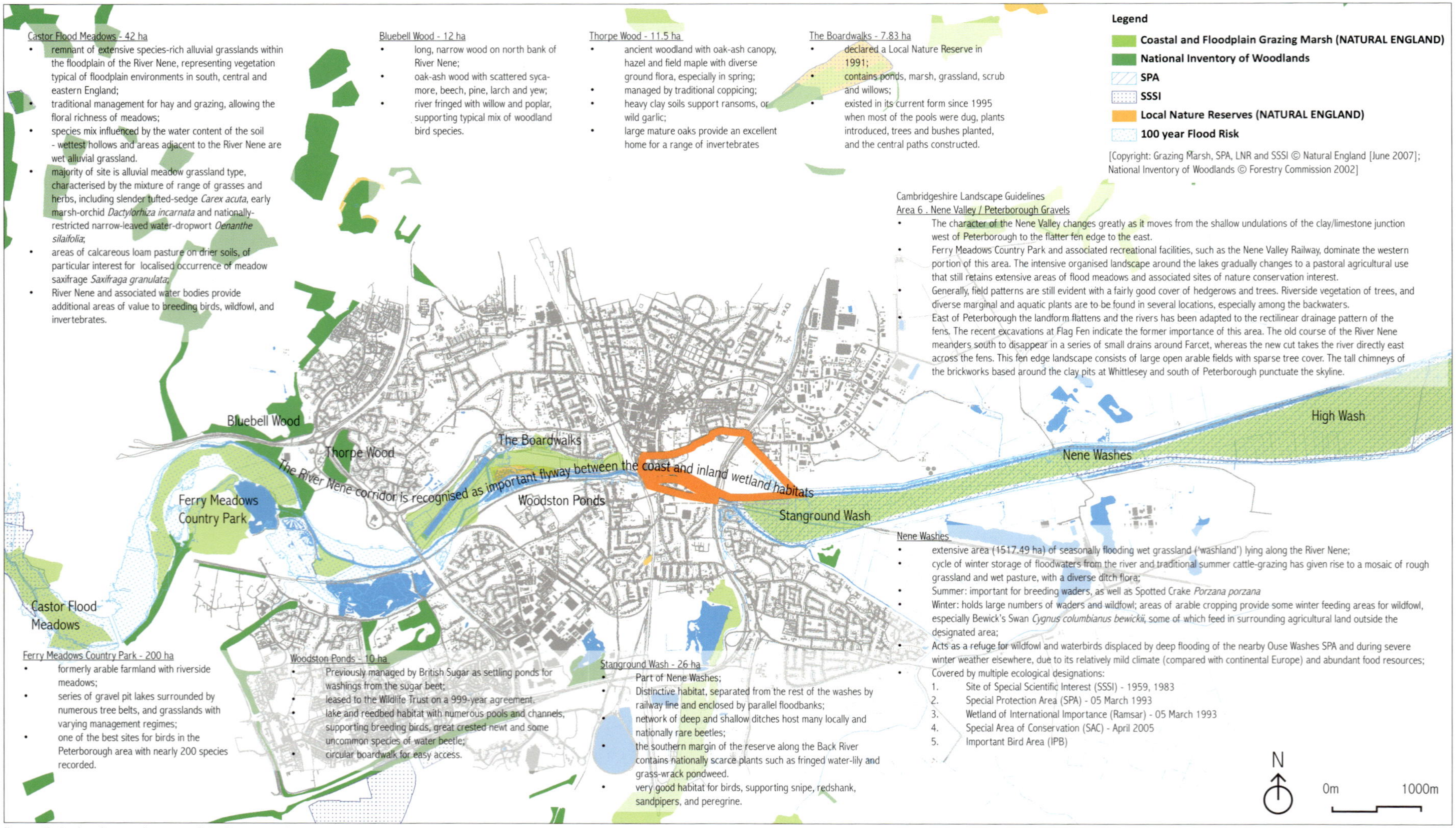

Figure B10 - Landscape character of the Nene corridor

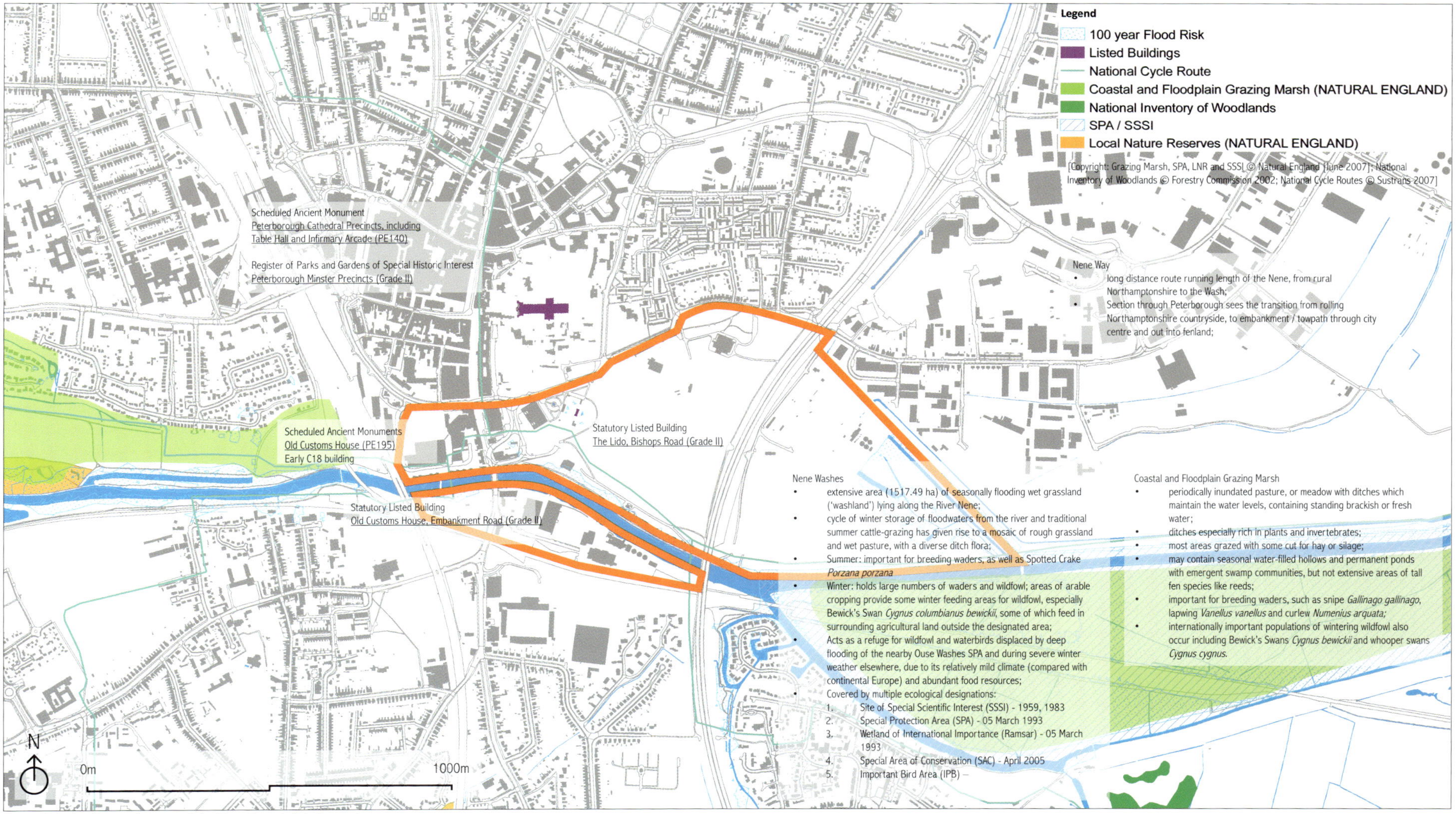

Figure B11 - Landscape character of the Nene through Peterborough

Flood Risk

Source

The main flood risk to the site is fluvial (river) flooding but in the future tidal flooding could become a consideration. Surface water flooding is not significant here and there are no reports of groundwater flooding.

There was a major flood on the River Nene in 1998, estimated in the order of a 1% probability event. Northampton was seriously flooded but Peterborough was little affected, with only a few areas flooded on both banks. Most notably these included the civic offices on the south bank and the new theatre set within in the park on the north bank. A large low-lying area within the Fengate site was also inundated. In 1947 another major flood inundated parts of the city including an area upstream of the development site and an area south of the site, in the Fletton/Stanground area.

The flood risk at Peterborough results from the buildup of the flood hydrograph through the upper and middle catchments of the River Nene (shown in figure B13). An important factor in this is floodplain storage, which leads to longer duration but lower peak flows as the flood moves downstream. Major development is ongoing in towns within the upper and middle catchment. There are implications for flood risk in Peterborough if this development is not accompanied by appropriate run off controls and flood hydrograph attenuation.

Extent

The Environment Agency flood risk map (figure B12) shows virtually all the site as being in the 1% probability flood risk area, but this assumes no flood defences and is based on outdated modelling.

There are flood defences and the evidence from the 1947 and 1998 floods suggests only a small part of the site is within the 1 in 100 year (1% probability) flood risk area, but that more of the site is in the 1 in 1000 year (0.1% probability) area. Ongoing modelling and mapping developed for the new Catchment Flood Management Plan (unpublished at time of writing) confirms this (figure B14). The modelling also indicates that the 1 in 20 year flood is contained within bank through Peterborough.

SFRA level 2 modelling was being produced (at the time of writing) which is likely to verify the Flood Zones to be used in assessing development.

	Flood Zone 1 (Low Probability) Areas with less than 1 in 1000 year (0.1%) annual probability of flooding.
	Flood Zone 2 (Medium Probability) Areas with between 1 in 1000 year (0.1%) and 1 in 100* year (1%) probability of flooding. * 1 in 200 year (0.5%) probability of flooding from the sea
	Flood Zone 3a (High Probability) Areas with between 1 in 100* year (1%) and 1 in 20 year (5%) probability of flooding. * 1 in 200 year (0.5%) probability of flooding from the sea
	Flood Zone 3b** (Functional Floodplain) Areas with greater than 1 in 20 year (5%) probability of flooding. ** This is not shown on the Environment Agency website.

Figure B12 - Environment Agency flood risk map

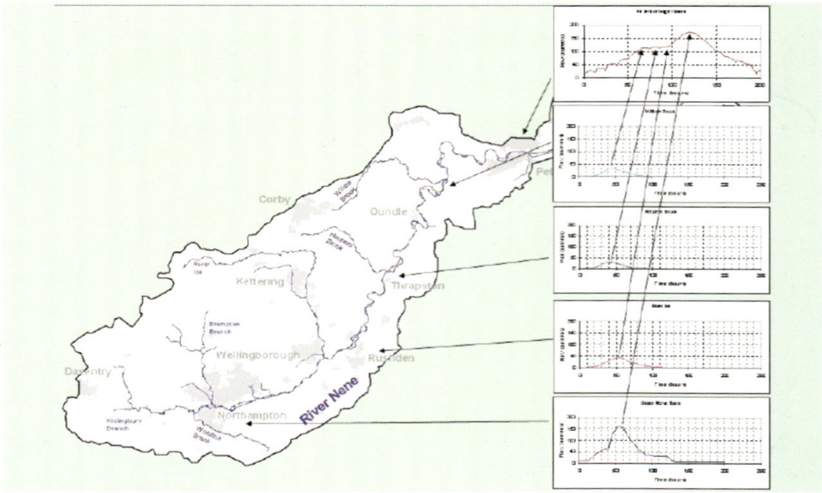

Figure B13 - Accretion of flows, from Catchment Flood Management Plan

For the purposes of this research the work was examined against the most up to date modelling made available. Figure B14 shows none of the development site to be within in the 1 in 25 year extent (similar to Flood Zone 3b) and only a small part, to the south east of the old sewage works within the 1 in 100 year extent (equivalent to Flood Zone 3a). Parts of the remainder of the site are shown within the 1 in 1000 year extent (equivalent to Flood Zone 2) and the outer fringes outside flood risk.

Duration and Depth

Fluvial flood risk is typically of long duration from the accretion of flows in the upper and middle catchment. Figure B15 shows the 1 in 100 year flood hydrograph downstream of the A1139 road crossing. It shows a maximum flood depth of 0.8 m, (close to the river) which decreases further away from the river. It also indicates a flood duration (water above bank level) at this point is excess of 150 hours (6.5 days).

The difference in water level between the 1 in 100 year and the 1 in 1000 year event was found to be approximately 1 m. Therefore land that is at risk of flooding from the 1 in 100 year event would flood to a significant depth during a 1 in 1000 year event.

Flood Risk

Flood risk in terms of the sources of flooding, the pathway or route by which the flood would occur and the receptor such as homes or farmland; the hazards for the site are set out in Table B1.

Surface Water

The EA does not currently provide flood maps for surface water, groundwater or sewer flooding. The EA provided no evidence of flooding from the land or sewers provided, however, as a precautionary approach it was assumed that flooding could potentially occur.

The proposed development areas at Peterborough have been previously developed to a large extent, i.e. they are 'brownfield'. As such the requirements with respect to surface water drainage may not be demanding, in that they may only be required to mimic surface water flows. Current EA policy is to aim for 'greenfield' equivalent infiltration and run off whenever possible. Therefore opportunities for improved attenuation and to prevent pollution reaching the watercourse should be considered.

The geology is thought to be alluvium over gravels (terrace deposits), which are appropriate for a soak away system. However, the groundwater table is likely to be high (controlled by the river), which could factor against an infiltration system.

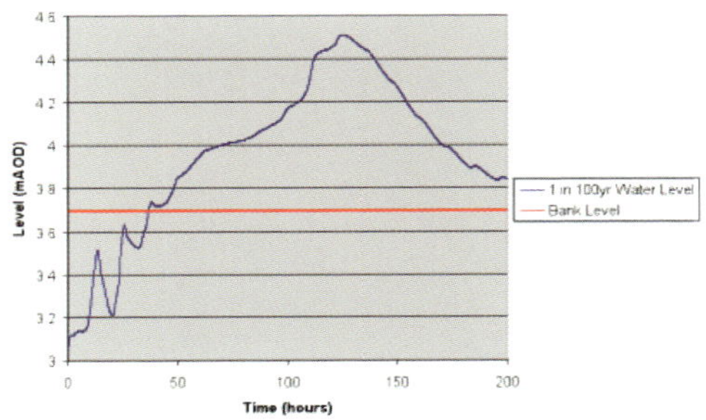

Figure B15 - Flood hydrograph for 1 in 100 year event

Figure B14 - Flood map, current (based on draft modelling data)

Site 2, Flood Risk		
Sources of Flooding	Pathway	Receptor
Flooding from Rivers	Overtopping of river banks	Public parkland
Flooding from Land (theoretical)	Run off from roads and land	City electricity substation
		Homes
Flooding from Groundwater (theoretical but no history)		Businesses
		Civic buildings

Table B1 - Flood risk to the site in Peterborough

Effects of Climate Change

Climate Change

In the future climate change will impact on hydrological conditions and flood flows are expected to increase for any particular frequency event. This will result in higher flood levels and longer duration for a particular frequency event.

Increased River Flows

Modelling information based on increased river flows of 20%, as set out in Table B.2 of PPS25, are shown in figure B16. This shows the same area of land within Flood Zone 3 but a far more extensive area of land within Flood Zone 2. The outer fringes on the north bank remain in Flood Zone 1.

Increased river flows greater than 20%* could result in an increase in the area of land at risk from flooding, as shown in figure B17. The extents appear similar to the Environment Agency flood map (figure B13). Therefore with climate change areas that currently fall within Flood Zone 1 could in the future lie within Flood Zone 3.

Sea Level Rise

Sea level rise could have a significant impact on the management of the Lower Nene. Though Peterborough is some 50 km (30 miles) inland from the sea it is low lying and without the Dog-in-the-Doublet sluice would be tidal. The east coast of England has a high tidal variation and is susceptible to storm surges; therefore if the Nene were to become tidal at Peterborough this would have a significant effect on the site. A number of sea level rise scenarios on the site were considered.

The current 0.05% high tide level at the Dog-in-the-Doublet sluice gate is +5.69 m Ordnance Datum Newlyn (ODN). With 1 m sea level rise (1045 mm is predicted over 100 years based on latest Defra guidance, October 2006) the 0.05% high tide level would be circa +6.6 m ODN, higher than parts of the site, requiring increased tidal control measures or resilient planning. High tides may also cause an increase in periods of tide lock, when the river level is below the tide level, and increase dependency on the use of the washes to store river water.

Sea level rise above 1 m is likely to require strategic decisions at a high level about the future of the Fens and areas outside the scope of this study.

With significantly higher rises in sea level it would be increasingly more costly and difficult to protect the Fens; Peterborough could become a coastal city, most probably with a heavy dependence on pumped drainage to cope with surface water.

For the purpose of this study, and transferability of ideas, the change to river flows formed the focus of the work and sea level rise was not considered further.

* Which could occur if periods of rainfall become prolonged in the future. Further urbanisation of the upper and middle catchments could also impact on the flood hydrograph.

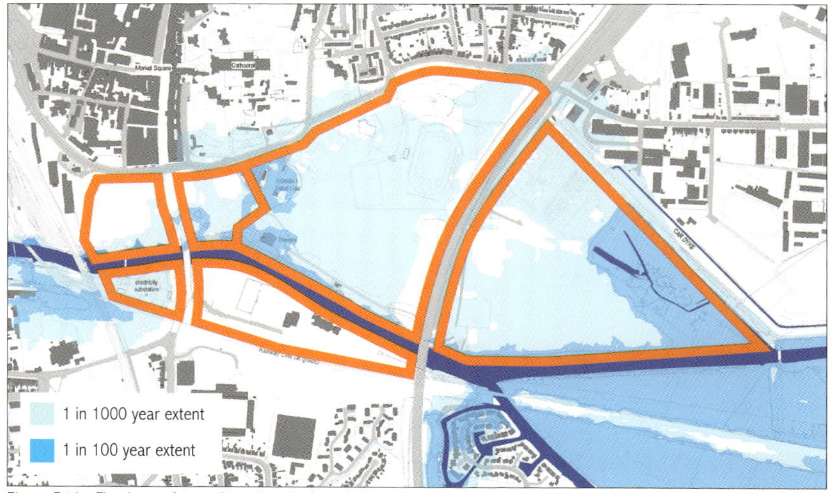

Figure B16 - Flood map, future (based on draft modelling information)

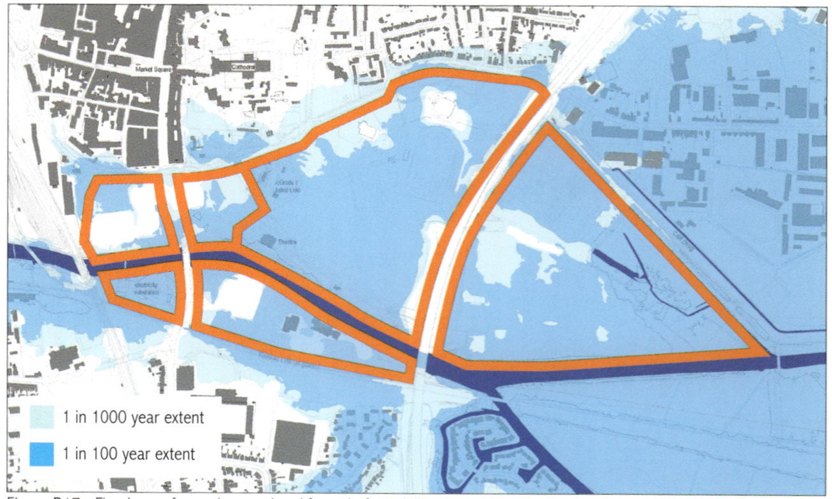

Figure B17 - Flood map, future (extrapolated from draft modelling levels to indicate sensitivity to increased flows and not verified by modelling or the EA)

Available Resources

Energy Sources

The available natural resources that would be suitable for energy generation at Peterborough are wind and solar. It is possible that a local aquifer could also be harnessed, particularly for cooling. Heat exchange with the ground would also be viable.

The fall across the entire site was not great enough to consider micro hydro generation, however this may be possible on steeper rivers.

The average wind speed is indicated as 5.7 m/s at 25 m above ground level and 6.2 m/s at 45 m above ground level. This would suggest that commercial 2 MW wind turbines and medium sized 150 kW wind turbines may be viable.

The land area is too small to grow a significant biomass crop on site. However, fuel could be provided from an off-site supplier.

Therefore the following technologies were identified as viable on this site:

- Wind
- Solar (Photovoltaic and Solar Hot Water)
- Biomass
- Ground Source Heat Pumps
- Aquifer Thermal Energy Storage (ATES)

Rainfall

The total site area (including central park) is 770,000 m^2 and the average annual rainfall is approximately 600 mm per year[44]. Therefore the total volume of rain falling on the site would be approximately 462,000 m^3 per year.

The 300,000 m^2 Fengate site would receive an average of 180,000 m^3 rain a year.

Formulating the Design Brief >

Development Mix

The accommodation schedule, shown in Table B2, was determined from past project experience and research into local demand. Using this schedule, energy demand, water demand and amenity standards could be determined.

A conceptual mixed use development of 2400 homes was proposed across the riverside sites in Peterborough, excluding the central park (50 hectares in total). A further 30,000 m^2 non-domestic buildings were also proposed.

The domestic property density would equate to an average density of 48 units per hectare, with higher densities of between 60 and 75 units per hectare closer to the city centre.

Amenity Requirement

Based on the accommodation schedule the number of inhabitants was estimated to be approximately 6200*.

An assessment of applicable national and local planning policy guidance was undertaken to determine minimum standards of open space required for the development.

The requirements for open space for approximately 6200 people were:

- Sports and play provision** – 13.45 Ha
- Outdoor (public) open space – 9.40 Ha***
- Private amenity space – 12.37 Ha
- Total required outdoor space – 35.22 Ha

Figure B19 illustrates the areas of outdoor open space required in more detail and the site area available.

* based on an occupancy of 60% of bed spaces

** Including formal sports pitch provision, children's play and other public sports areas

*** Including 0.63 Ha allocation for allotments

Site 2, Accommodation Schedule

Residential	GIFA (in m²)	No. of Homes	Total Area (inc circulation)	Parking
1 bed studio	37	0	0	0
1 bed flats	50	610	35,075	610
2 bed flats	70	610	49,105	610
3 bed flats	90	360	37,260	720
3 bed houses	100	240	24,000	480
4 bed houses	120	360	43,200	720
5 bed houses	140	240	33,600	480
6 bed houses	160	0	0	0
Total Residential GIFA (in m²)		2,420	222,240	3,620
Commercial			GIFA (in m²)	Parking
Office Area			8,256	413
Retail Area			16,512	826
Industrial Area			2,064	103
Educational Area (inc nurseries, creche..)			2,064	103
Health and Community Facilities			2,064	103
Total Commercial GIFA (in m²)			30,960	1,548
TOTAL GIFA			253,200	5,168

Table B2 - Accommodation schedule

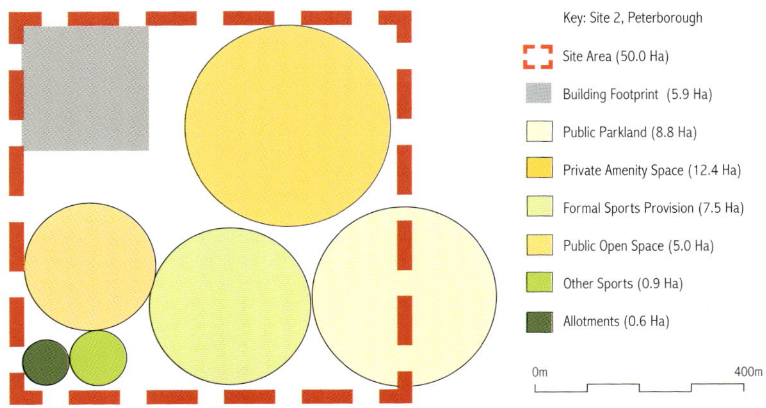

Figure B19 - Spatial requirements for open play

Resources

Energy Demand

Using a range of typical energy consumption figures for different building types (details included in appendix D) the accommodation schedule was used to calculate the total energy demand for the Riverside sites, indicated in Table B3 and figure B20.

Water Demand

The total water demand was calculated as 757,300 litres per day, 238,250 m^3 per year (based on 110 l/p/day for residential, which incorporates dual flush WCs and low flow taps + 11 l/p/yr for commercial).

This water demand is approximately 50% of the total rainfall across all 77 hectares of the site.

Site 2, Energy Demand						
Energy Demand (MWh/yr)	Space heating	Space cooling	Hot water	Electricity (excl cooling)	Total (MWh/yr)	Percentage
Total Residential Demand	3,095	0	6,720	8,690	18,505	74.5%
Total Commercial Demand	750	1,505	360	3,735	6,350	25.5%
Total Energy Demand	3,845	1,505	7,080	12,425	24,855	100.0%

Table B3 - Energy demand

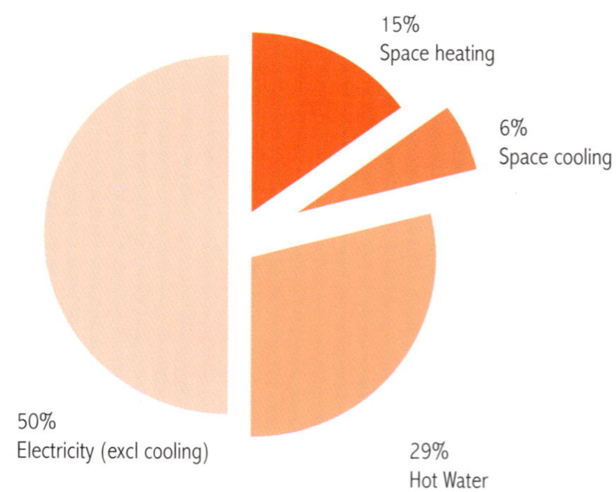

Figure B20 - Energy demand mix

SUDS Requirement

The design standard for surface water drainage is 1 in 30 years, ie that the capacity of the drains can cope with a severe rainstorm, only likely once every 30 years. However, the design standard for flooding is 1 in 100 years. During a 1 in 100 year rainstorm drains are likely to be overwhelmed.

Current EA policy is to aim for 'greenfield' equivalent infiltration and run off whenever possible.

This design standard is adopted by the LifE project: to provide surface water drainage that can cope with a 1 in 100 year event plus a 30% increase in peak rainfall to account for climate change to 2115.

The sub-strata is thought to be alluvium over gravel, which would have a high permeability, however, the groundwater table is likely to be high, reducing the ability for the water to soak away. It is also likely that the former industrial sites and landfill will contain some contamination and therefore infiltration will not be an option and a sealed system would be required.

A precautionary assessment for the volume of attenuation would be 300 m^3 rainfall per hectare*. This would equate to a total attenuation volume of 23,100 m^3 across the entire site, approximately 14,100 m^3 for the developable site and 9,000 m^3 for the Fengate site. With an average swale depth of 0.5 m an area of 18,000 m^2 attenuation within the Fengate site would be required.

The total optimum water requirements are shown in figure B21.

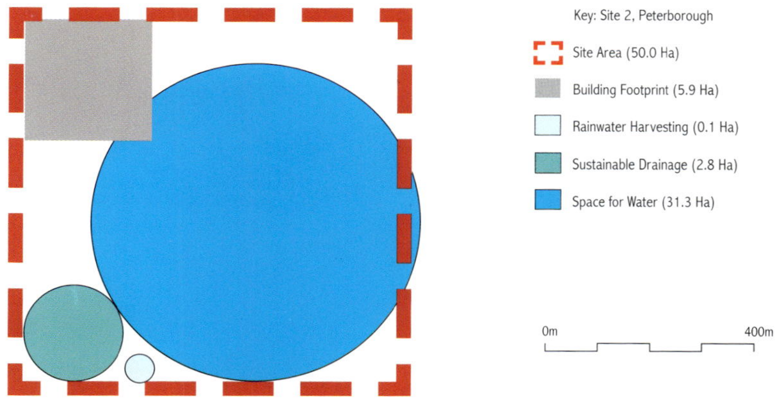

Figure B21 - Spatial requirements for rainwater, attenuation, harvesting etc..

* 300 m^3 per hectare equates to 8 l/s/ha for a 10 hour rainstorm or 5 l/s/ha for a 16 hour rainstorm.

Design Objectives >

The objective was to establish a strategic plan for the whole riverside site that would enhance the existing assets and characteristics of the city and improve access to the river.

Flexible and resilient buildings would be set within a resilient landscape to cope with high volume water inundation, aid fast recovery, and minimise economic impact of a flood.

Design Objectives

Based on the available resources, the land available in Flood Zones 1 and 2, and the requirement for amenity space, the following strategy was developed to enable a suitable development to come forward.

Main Objective

Create a landscape-rich development that provides for the multi-functional amenity requirements of a medium density suburban population and simultaneously reduces surface water run off and flood risk to lives and property.

Landscape Approach

Develop property in high density clusters to minimise development footprint and preserve land for future adaptation.

Enhancement of the river corridor for waterfront and cultural activities would integrate with the city's green infrastructure network, provide flood and water management, opportunity for local food and energy and reconnect the town to the river.

Energy and Water Approach

The flat expanse of the fens and river corridor has the potential to provide a considerable proportion of the energy demand from large-scale wind turbines, not only to the site but the rest of Peterborough. Preservation of views of the cathedral and the 'big skies' will restrict this opportunity; therefore solutions that integrate wind power within developments will be preferable.

Aim to provide at least 50% of water requirements for flushing WCs and irrigation from rainwater harvesting to reduce the water consumption to 80 litres per person per day.

Flood Risk Management Approach

The volume of water resulting from the flood hydrographs precludes consideration of storage over and above the capacity of the floodplain as a means of flood risk management. Defending the site, increasing the river channel capacity or diversion of flood flows, are also not appropriate.

For this site effective flood risk management options are limited to working with the risk and planning the development accordingly. Consideration of flood storage compensation and conveyance related to development within the floodplain will be necessary.

And to manage all surface water run off for a 1 in 100 year storm event using sustainable drainage.

Concept Ideas

The concept plan (figure B22) shows new development wrapping around Central Park, orientated towards the Cathedral and the river. The main concepts of the plan were:

- Develop along the main road into Peterborough to re-establish a gateway into the historic city centre.
- Enhance the visual and physical connection with the Cathedral and the City Centre by organising development around vistas and routes.
- Prioritise higher density development close to the city centre and in Flood Zone 1.
- Arrange development around flood flow pathways to allow water to pass through the development with minimum harm and disruption.
- Integrate wind turbines within green corridors to reduce their visual presence.
- Transform Central Park into a high quality multi-use civic park that is designed to cope with occasional flooding.
- Create a riverfront corridor for recreation and amenity value and to better serve as a corridor for wildlife moving between habitat up and down the River Nene.
- In areas of greatest uncertainty, establish an adaptable masterplan, which can allow development of buildings and infrastructure now and their removal or change of use in the future to reduce vulnerability.

Four options (figure B23) were considered for the development of the Fengate site. The existing rail and road embankments create barriers to flood flows. Development could therefore be considered adjacent to the river, where the risk from flooding is lower than would normally be expected in this situation. 'Development Islands' were selected on the basis of creating a logical flood risk management plan, with flexibility for future adaptation.

Development Islands

Narrow islands of development would run parallel to the river, bisected by drainage corridors. Conventional street block arrangements were inverted such that parking and access were located within the centre, instead of gardens.

Back gardens would front onto the semi-public rain corridors and secondary access routes. These would provide pathways for flood flows and wind to drive turbines.

Safe access would be provided by roads elevated above the drainage corridors but below the maximum flood level to preserve floodplain. Home zones would be located above the flood level, preventing risk of damage from vehicular wash during a flood.

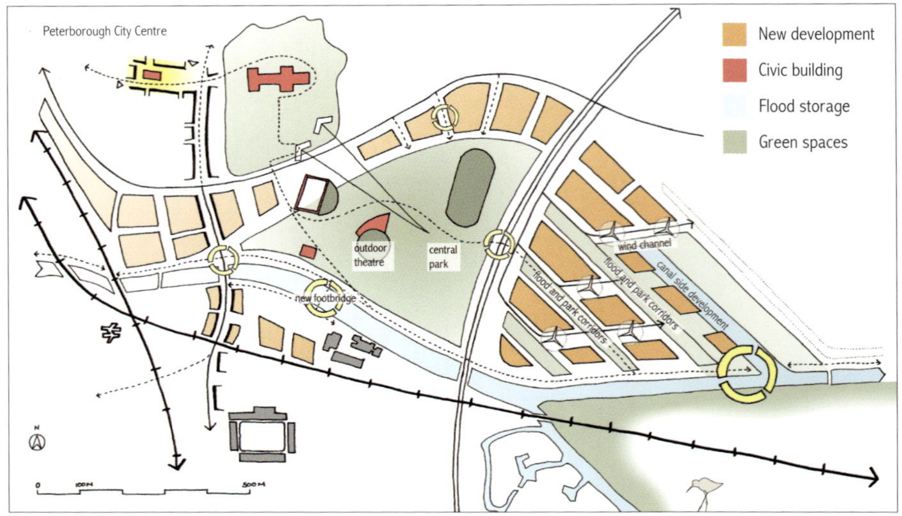

Figure B22 - Concept plan

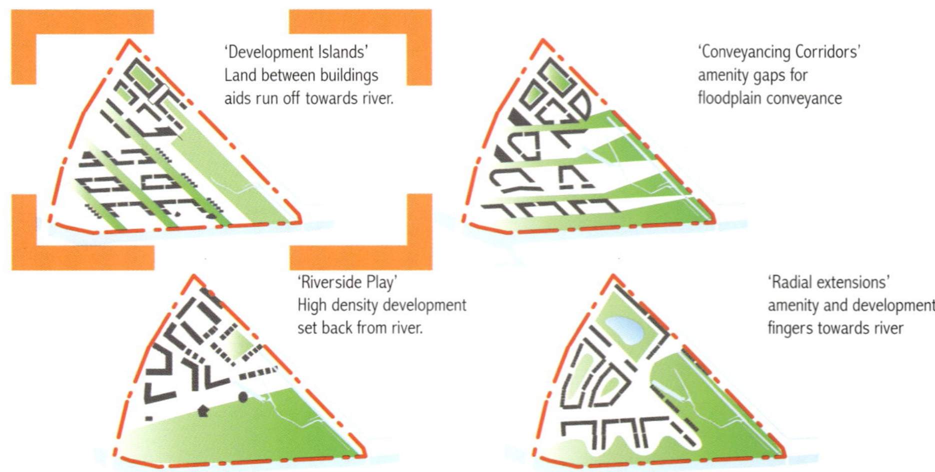

Figure B23 - Plan options

A conceptual masterplan was developed for the Fengate site that evolved around the flood flow paths, in this case drainage paths to allow flood water on and off the site as quickly as possible.

This is illustrated in figure B24.

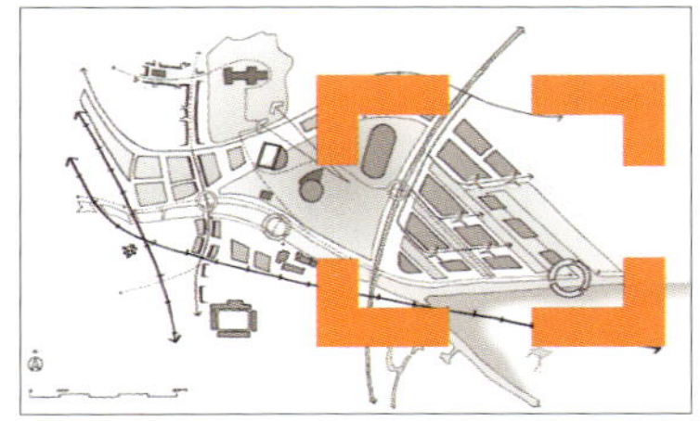

Stream Corridor – Development is organised around secondary movement corridors that are used primarily for amenity and pedestrian access. The ground is excavated down to the water table in places to maximise floodplain compensation and to create a permanent winter water body for wildlife.

Resilient Play – Play areas are located in higher flood risk areas. Resilient materials are designed to be cleaned easily to allow a fast recovery after a flood. Advance warning times should suffice to ensure no occupants are present when a flood occurs.

Adaptable Buildings – Buildings closest to the river are designed with wide plans and high ceilings to lower floors to allow future conversion from residential to commercial if required.

Energy Corridor – The stream and rain corridors also provide the opportunity for wind energy generation. The wider corridors provide generous space between buildings for light and separation from the noise of turbines.

Figure B24 - Conceptual masterplan for Peterborough Fengate site.

Green Shelf – A generous open space set below the level of development provides numerous community and recreational activities, both quiet and active, from organised sports to allotment gardens. Wider corridors offer opportunities for more uses.

Home Zone - A raised home zone remains dry in extreme flood events providing a safe place and access route. The inclusion of gravel swales, tree planters and porous paving all provide additional rainwater attenuation.

Rain Corridor – At higher levels on the site amenity corridors are organised around wet swales and attenuation ponds to reduce the run off down into the higher flood risk areas. These corridors (conceptually linear rain gardens) are integrated with pedestrian and cycle paths to provide an attractive feature and an important drainage corridor.

Amphibious Buildings – Selective buildings designed to rise from their foundations during time of flood could be located closer to the water course.

Energy and Resources

The opportunity to use wind power, especially larger turbines on site was appealing, due to the potential capital savings over other technologies and improved likelihood of meeting all of the energy needs from on-site technology. Therefore three energy options were explored, using different scales of wind turbines.

— Option 1 - Small-scale wind power and solar PVs

— Option 2 - Medium-scale wind power and solar PVs

— Option 3 - Large-scale wind power and solar PVs (option ruled out)

Option 1 would require all available roof area for PVs to minimise the number of turbines required. Using 57,170 m^2 PVs left a requirement for 128 15 kW turbines.

Use of the larger wind turbines reduced the requirement for PVs. However, planning obligations to maintain the views of the Cathedral in combination with other siting constraints are likely to rule out the large-scale wind turbines and potentially even the medium-scale turbines. Therefore additional roof area such as over the sports hall, and use of less efficient wall-mounted solar panels were considered (Table B4).

The resulting energy solution was primarily based on biomass CHP, with top-up biomass boilers providing approximately 45% of the total energy demand. The energy technologies in order of contribution were:

— Biomass CHP (fuel from off-site supplier)

— Solar PVs

— Wind Turbines

— Aquifer Thermal Energy Storage (ATES)

— Solar Thermal

— A small contribution from Biogas

A final solution would aim to use a mix of 15 kW and 150 kW turbines to reduce the number of turbines.

Access to the aquifer provides a significant contribution to the heating demand (approximately 1/3 total demand), thus reducing the dependency on biomass for heating. However, the heat pumps require additional electricity to run, increasing the power necessary from Solar PVs or Wind.

* Note that these figures have been rounded to the nearest 5 MWh

** For total biomass demand, area of willow coppice required to be self sufficient in biomass = 177 hectares (or 0.08 hectares per residential property).

*** Potential to also produce biogas from sewage sludge, but would only add extra 6 MWh/yr so not deemed cost effective for this population. Potential to widen area of catchment for both kitchen waste and sewage sludge. However, at present this is not included for as it would be using another population's resource for the purposes of making this particular site Zero Carbon.

**** Assuming flat plate collectors

***** The total roof area for the development is around 90,000 m^2. Not all of this area will be available for solar panels (solar water heating and PV). Assuming 70% would be suitable for panels (this may also include some suitably orientated façade areas), total available area = 63,000 m^2. Assuming 2,970 m^2 roof area required for solar hot water (assuming 10% for frame etc), this leaves 60,030 m^2 for PV, equivalent to around 57,170 m^2 active PV panel area (assuming 5% for frame etc).

Option 3 was subsequently ruled out due to concerns over migratory birds using river corridor, and hence increased chance of bird strikes to large turbines. Option 1 is probably impractical due to large number of turbines required. Therefore most pragmatic option appeared to be Option 2 where 13 medium scale 150 kW turbines are used in conjunction with maximum PV panel area on the site.

Technology	Yearly energy demand met*	Which demands met	Further info	% of total
Biogas CHP	50 MWh 35 MWh	HOT WATER: max output based on biogas production capabilities ELECTRICITY: small contribution	For the population of the site (approx 7,050), biogas production using anaerobic digester utilising kitchen waste, max biogas production = approx 110 mWh/yr ***.	0.4%
Biomass CHP	4,820 MWh 1,335 MWh 2,800 MWh	HOT WATER: Balance of energy required, after biogas CHP and solar thermal panel hot water production SPACE HEATING: 50% of residential demand ELECTRICITY	Approx 0.5 MWe biomass CHP required. Annual fuel use approx 13,900 m^3 wood chip, or 3,450 m^3 wood pellet **	41.1%
Biomass boilers	810 MWh	SPACE HEATING: Balance of energy required after biomass CHP and ATES production capabilities	Annual fuel use approx 1,200 m^3 wood chip, or 300 m^3 wood pellet **	3.7%
Solar water heating	1,220 MWh	HOT WATER: 50% of all houses	Approx 2,700 m^2 total active panel area required****	5.6%
Inter-seasonal thermal storage	1,170 MWh 1,170 MWh	SPACE COOLING: 90% of total SPACE HEATING: for thermal balance of system	Aquifer thermal energy storage (ATES) assumed. Sizing for peak demand is inefficient, therefore 90% of yearly cooling demand only. For max efficiency, system should be in thermal balance.	10.7%
Three options considered for further onsite electricity production:				
Option 1 PV (max available)	5,715 MWh	ELECTRICITY: To run ATES; to run chiller to satisfy remainder of cooling demand; elec output from largest PV installation possible onsite	Approx 57,170 m^2 total active panel area *****	26.2%
Option 1 Small scale wind	2,545 MWh	ELECTRICITY: To make up balance of demand	Approx 128, 15 kW turbines required	11.7%
Option 2 PV (max available)	5,715 MWh	ELECTRICITY: To run ATES; to run chiller to satisfy remainder of cooling demand; elec output from largest PV installation possible onsite	Approx 57,170 m^2 total active panel area *****	26.8%
Option 2 Medium scale wind	2,545 MWh	ELECTRICITY: To make up balance of demand	Approx 13, 150 kW turbines required	11.7%
Option 3 Large scale wind	10,500 MWh	ELECTRICITY: To run ATES; to run chiller to satisfy remainder of cooling demand, to make up balance of demand	Approx 3, 2 MW turbines required	38.5%

Table B4 - Proposed energy solution

Figure B25 illustrates the proportional contribution of each energy technology. Figure B26 indicates the spatial demands of the energy technologies, including the amount of land needed to grow biomass to support the annual fuel demand.

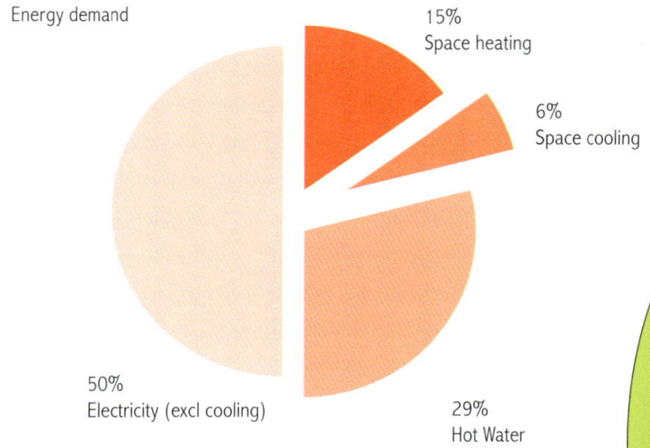

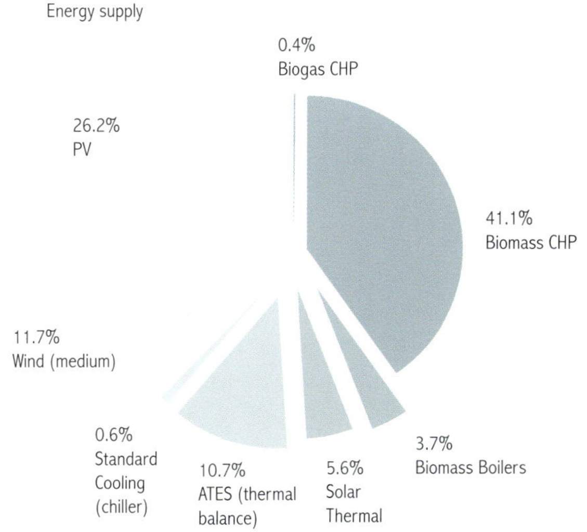

Figure B25 - Energy technology contribution to meet total supply (option 2)

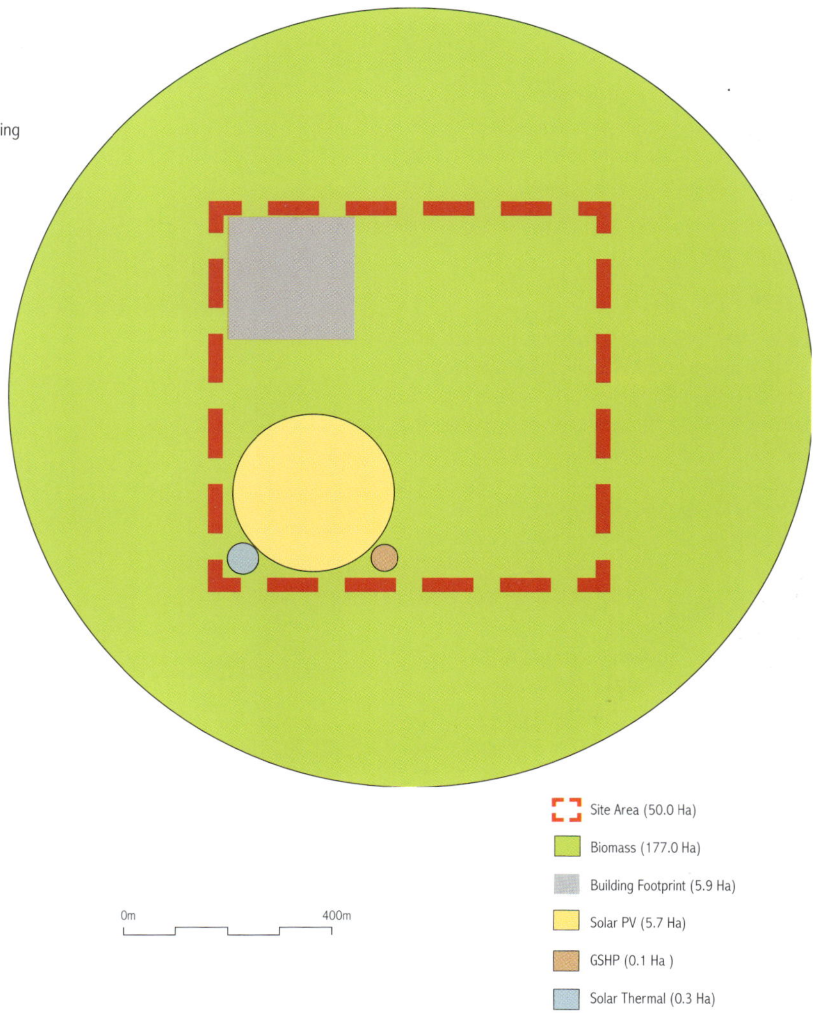

Figure B26 - Space requirements for energy technologies

Strategic Energy Ideas (outside study area)

The Peterborough Energy Corridor

The flat expanse of the fens and river corridor has the potential to provide a considerable proportion of the energy demand not only to the site but the rest of Peterborough (figure B28).

Figures B27a to B27c indicate proven wind turbine technologies.

Large Wind

Geographically it might be possible to install 40 2 MW turbines from Peterborough to the Ring's end sluice (see figure B29). This would equate to a total of 140,000 MWh/yr. However, in practice it is unlikely that this number could be supported in this location. Ten turbines could provide in the order of 35,000 MWh/yr (equivalent to 4000 homes total electricity demand) and more may be possible across the low lands.

Medium Wind

Eighty five 150 kW turbines (figure B27b) would provide in the order of 17,100 MWh/yr (equivalent to 2000 homes total electricity demand).

Biomass

This concept of an energy corridor could also see the river used to transport biomass from agricultural land downstream. This could therefore be used to support a wider energy vision for Peterborough.

Figure B27a - 2 MW wind turbines
Green Park Reading (left) and Ford, Dagenham (right)

Figure B27b - 150 kW wind turbine, Japan

Figure B27c - 15kW wind turbine

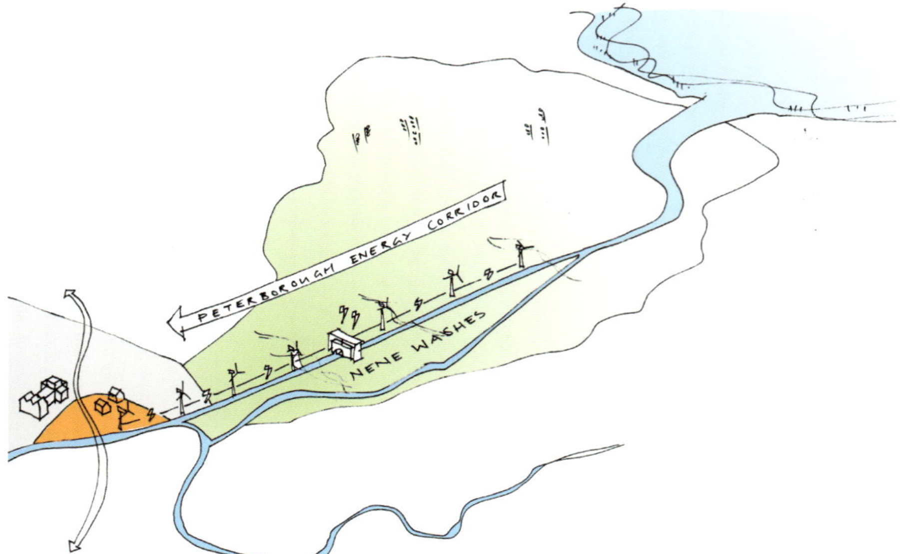

Figure B28 - Peterborough energy corridor

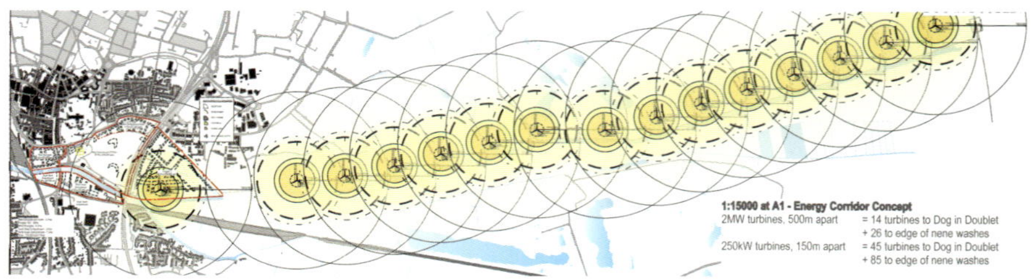

Figure B29 - Plan study of the minimum spacing required for wind turbines

The Water Solution

To reduce the water consumption from 110 litres per person per day (the level of consumption set by the Code Level 3) to the target 80 litres per person per day set by the Code Level 6, a further saving of 30 litres per person per day would need to be provided from rainwater harvesting.

If the number of consecutive dry days is likely to be a maximum of say 20 the total required harvesting would be approximately 3,720 m^3. If it is only five days then 930 m^3 storage must be provided.

On the basis of five days, 25 number 40 m^3 tanks would be required within the Fengate site to achieve sufficient capacity to meet Code Level 6.

Figure B30 shows a communal rainwater harvesting tank located within each home zone to serve the needs of the surrounding occupants. These would be between 40 m^3 and 50 m^3 and connected to the gravel swale SUDS system. These tanks would supply over half the local water flushing and irrigation requirements, with surplus storage allowing capacity during flooding.

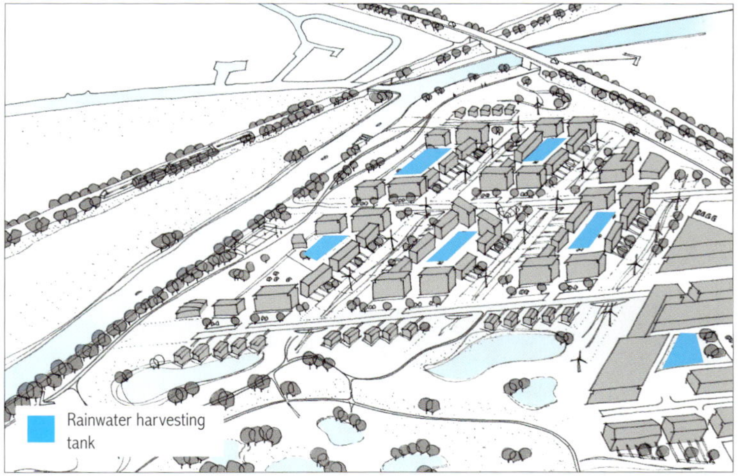

Figure B30 - Rainwater harvesting areas

Flood Risk and Development

One of the first principles of PPS25 is to discourage development in flood prone areas. A sequential test should be used to identify if alternative sites within Flood Zone 1 are available, particularly for more vulnerable development such as residential. This has not yet been completed (at the time of writing) for the riverside sites in Peterborough, but there is local support to redevelop the former industrial areas and potentially the Fengate site.

Based on avoidance as the principal approach to flood risk management, set out in the PPS25 practice guide[36], the suitability of the sites for residential development was not established. If the sites were to be considered for development it is likely that these would require less vulnerable development or mixed use.

Therefore the options considered focus on identifying appropriate development, based on the PPS25 guidelines set out in Table D.3[9]. The main approach was to allocate development to areas of lowest probability of flooding. Further management measures would include: compensation for loss of floodplain, consideration of flood conveyance, and site drainage of receding floodwater.

Figure B12 showed that the majority of the site areas were within Flood Zone 1 and therefore suitable for all types of development. Figure B16 showed that these areas would mostly remain within Flood Zone 1 after a 20% increase in flows from climate change. The Fengate site was found to be the most susceptible to climate change therefore this is the focus of this section.

Fengate Site

Figure B31 indicates the flood extents to the Fengate site, based on the information available. The site, which appears to lie predominantly within Flood Zone 1, therefore in accordance with Tables D.2 and D.3 of PPS25 all forms of development would be appropriate. However, the SFRA level 1 information contradicts this mapping and without the more detailed SFRA level 2 information it is likely that the EA would take a more precautionary view and revert to the worse scenario. In addition the effects of climate change, including future tidal influences resulting from rising sea levels, could increase the area at risk of flooding.

A precautionary design approach was used in which the majority of the site was assumed to be Flood Zone 2. On this basis residential use could be acceptable, subject to managing risk, but that Highly Vulnerable uses, such as emergency services, would not.

The proposal for a mixed use development or office based development, with generous open areas could be more appropriate for the site. Careful consideration was given to providing safe access to the site, above the predicted flood level, via elevated primary vehicle routes. In addition compensation for future loss of floodplain was calculated and designs were developed that would manage the remaining flood risk without impacting on other sites.

Figure B32 indicates the sources of flooding, showing the possibility of surface water flooding in lower areas of the site.

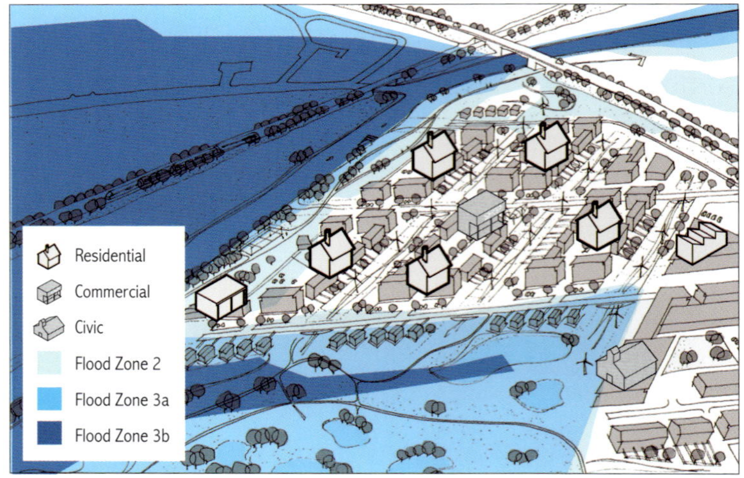

Figure B31 - Development and flood risk

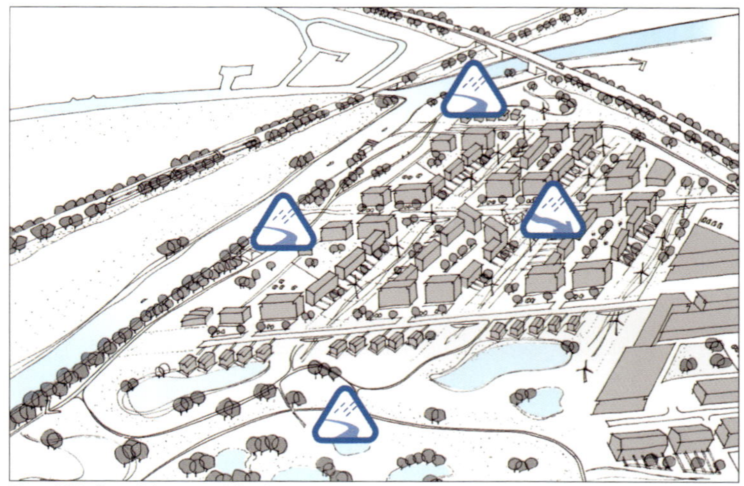
Figure B32 - Potential sources of flooding

Floodplain Compensation

Building within the floodplain will reduce the storage volume that is naturally available to the river in times of flood, with the potential to increase flood risk upstream or downstream of the site. To compensate for this it is necessary to provide an equivalent storage volume, which should be created as an extension to the natural floodplain. Ideally this compensatory storage volume should be provided on a level for level equivalent of what is removed for development. The compensatory storage volumes were calculated as:

- Zone 3 (1% or 100 year flood) = 6,875 m^3 between 3.5m and 4.5 m AOD
- Zone 2 (0.1% or 1000 year flood) = 38,850 m^3 between 3.5m and 5.5 m AOD.

On the basis that Zone 2 could become Zone 3 in the future a precautionary approach would be to provide 40,000 m^3 compensatory storage. This was proposed to be provided within the development itself, raising areas within Flood Zone 1 and lowering areas between buildings. This approach would need to be modelled to prove that flooding did not increase elsewhere.

Floodplain Conveyance

The A1139 road crosses the river and railway over a long-span bridge and is embanked across the floodplain. The only opening through the embankment is a pedestrian/cycle subway about 300 m north of the river and so the embankment forms a barrier to flow (conveyance) along the floodplain. Car Dyke is also raised and limits conveyance of water that spills onto the floodplain downstream of the road bridge. The land proposed for development acts as a floodplain reservoir and is therefore passive, rather than active, in terms of floodplain designation.

The main issue is therefore site drainage as flood levels in the river recede. This is incorporated within the site layout with the overall drainage direction being towards the southeast corner of the site, as shown within figure B33.

Surface Water Drainage - SUDS

PVs have been installed on green roofs in some developments but this practice is not widely used. Therefore rainwater attenuation would be provided within home zones in land above the floodplain; and additional storage would be provided within the rain corridors, between buildings.

The low frequencies with which the site will be flooded from the river make it feasible (subject to detailed investigation) to combine floodplain drainage paths and site surface drainage. This could help further environmental enhancement of the drainage corridors.

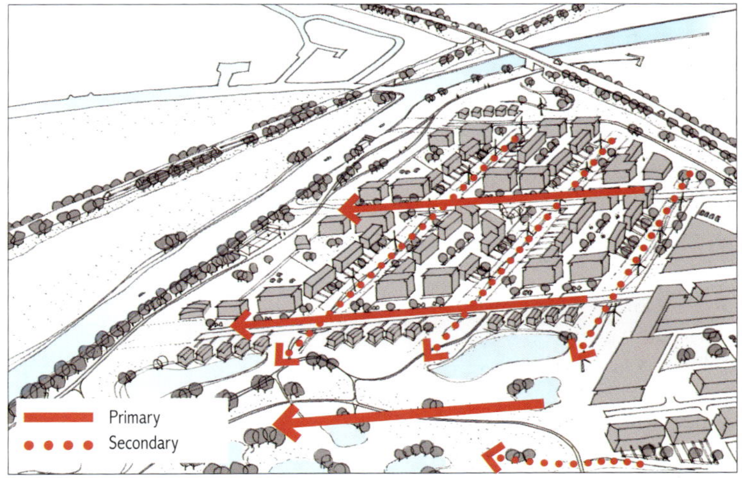

Figure B33 - Site drainage

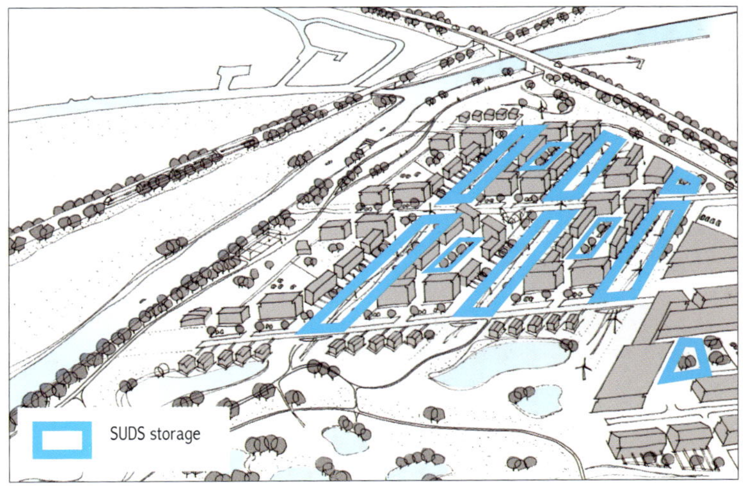

Figure B34 - Rainwater attenuation

Adaptation

Safe Access and Egress

Main access routes into the site would be located on higher ground and run perpendicular to the river. Main footpaths would be approximately 300 mm above the roads and accessed across wheelchair compliant ramped parking areas (see figure B35).

The wide strips of land between the development islands would provide secondary access routes and surface drainage, see figure B36.

By reducing car access into the lower regions of the site and favouring pedestrian and cycle movement, multiple lines of conveyance and drainage can be established.

Adaptation

The potential impact of climate change could be considered when designating the use and detail design of buildings on the site.

Buildings in areas of highest variability would be designed to be resilient to flooding, even though the risk may never materialise.

Buildings could be constructed with high ceilings and wide floor spans to allow conversion from residential to commercial if future extremes become more normal.

Figure B37 illustrates how areas that could be most affected by climate change could be designed for future removal of buildings or change of use.

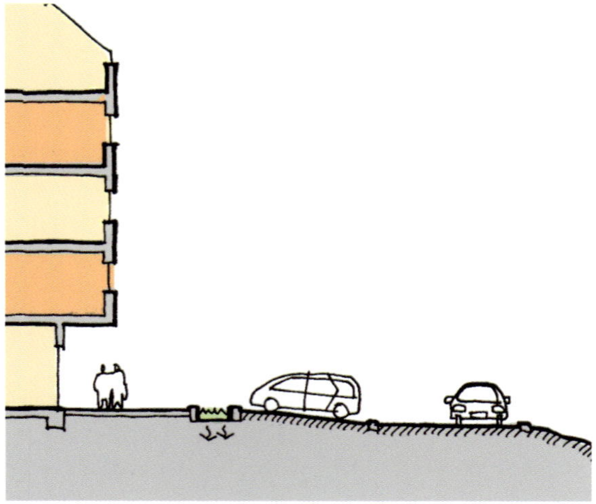

Figure B35 - Terraced access routes

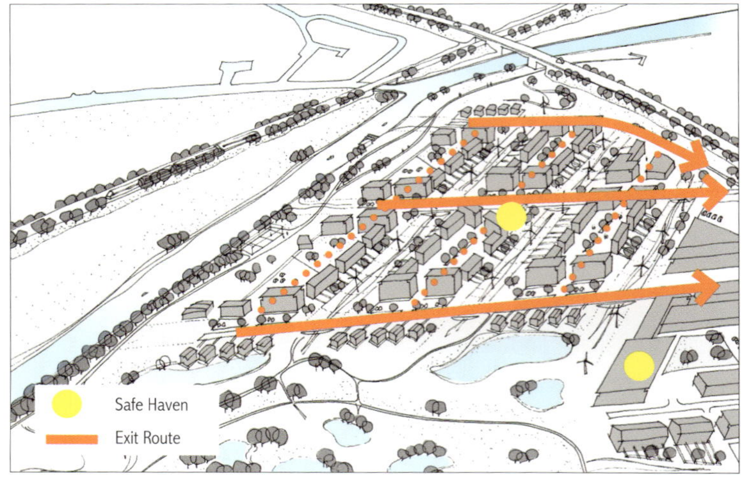

Figure B36 - Safe havens and exit routes

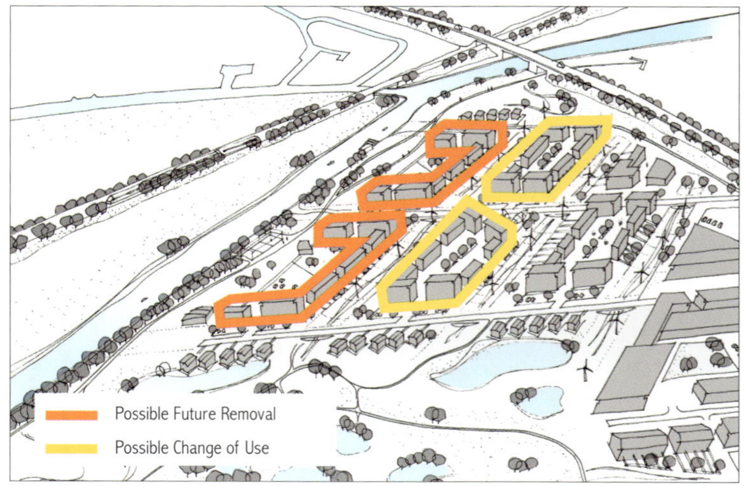

Figure B37 - Rainwater attenuation

Design of development in narrow parallel strips set back from the river not only allows generous areas for flood drainage but also creates a terraced street pattern that closely follows the natural topography.

Appropriate building technologies could be selected on the basis of flood risk, i.e. more adaptable building types for higher risk in areas of higher potential inundation.

Figure B38 shows a cross section through the Fengate site. Section A shows development across the site up to the current extent of Flood Zone 2, near the river. Section B shows how buildings closest to the river can adapt to higher inundation levels. Section C shows that if the site becomes susceptible to flooding far more frequently then the land can be changed to floodplain.

There are likely to be many obstacles to achieving land sales and development on the basis that a site may be made floodplain in the future. This approach would likely favour leasehold sales or short-term license for use.

In reality once developed it would probably be cheaper to identify flood storage upstream. However, this would allow development to occur in flood risk sites where no alternative sites exist, but still provide the mechanism to recover the land in the interest of flood management.

This approach could help to provide flexibility in meeting development pressure and managing flood risk.

Figure B38 - Sections

Landscape

The total amenity requirement for a population of 6200 people would easily be met by the city 'Central Park' which is within the extents of the study area and walking distance from each of the sub-sites. The park has large informal play areas and good sports facilities. It would be expected that a contribution from the developer would be required towards the upgrade of these facilities.

In addition, local play areas and outdoor areas of play would be provided on site. These are closely linked to the development areas, located within the stream and rain corridors, shown in figure B39.

Native trees and herbaceous planting would be introduced within corridors (figures B40 and B41) and the streetscape with water-loving varieties such as willows and poplar located in the wetter corridor areas.

Along the river, stream edges and standing bodies of water, reed beds and marginal vegetation would be established. Areas of grass would be subject to varied mowing regimes to create a mosaic of habitat for birds and invertebrates.

Figure B40 - Pollarded trees, characteristic of fenland landscapes help support biodiversity

Figure B41 - Water meadows within the centre of Cambridge

Figure B39 - Areas for outdoor play

Stream Corridor

A wet swale with wetland planting and attenutation ponds could provide an attractive feature and an important drainage corridor. An example of how this could be arranged is shown in figures B42 and B43.

Allotments and local play areas would be provided to the rear of back gardens.

Corridors would be sized to provide acoustic and vibration set back from wind turbines.

Precedents

Figures B44a to B44d show examples of other sites where similar ideas have been implemented.

Plans and Sections

Figure B45 shows an indicative development plan for the Fengate site. The home zones are designed to be narrower than the rain and stream corridors to create more space for water. Figures B46a to B46d show a section through this plan.

Figures B47a to B47c illustrate an area of the Fengate site before and during storms.

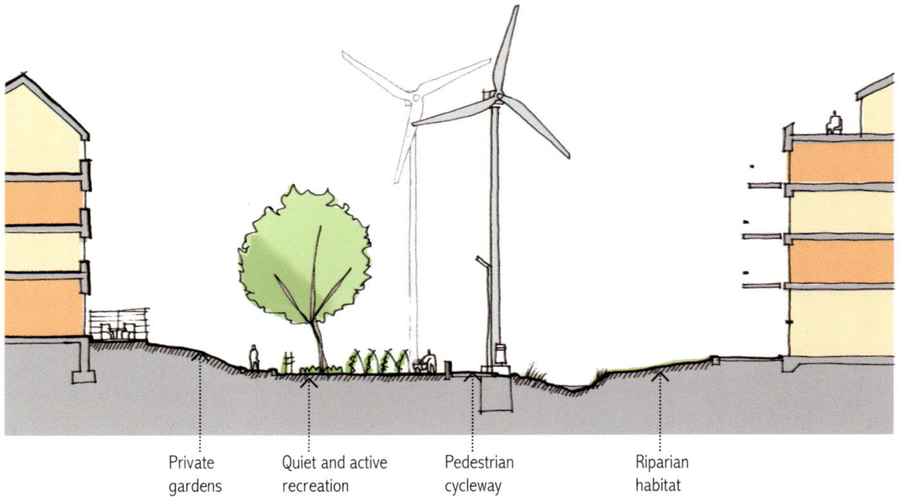

Figure B42 - Section through a stream corridor

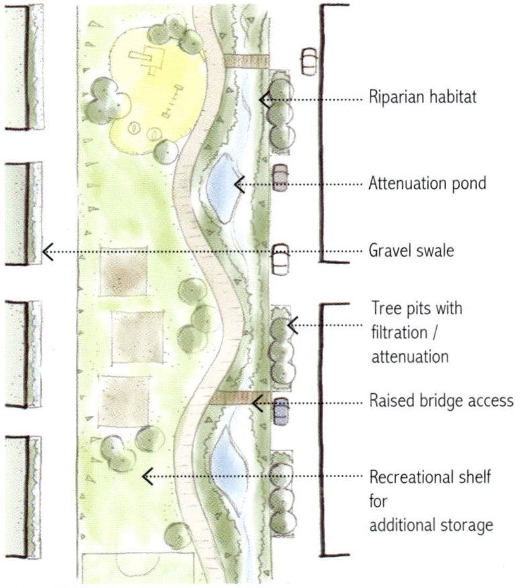

Figure B43a - Plan of stream corridor

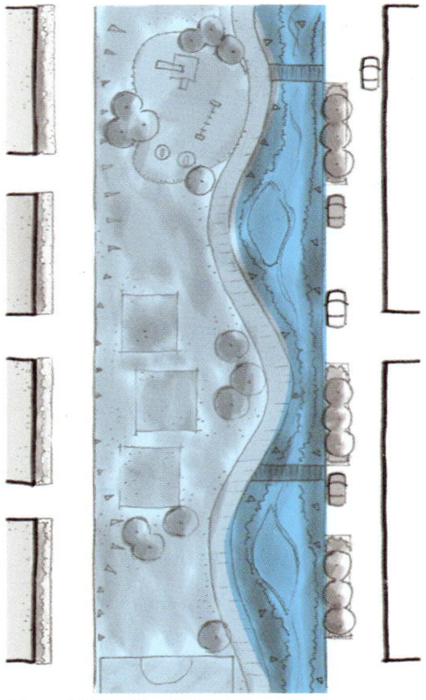

Figure B43b - Flood storage area

Precedents

Figure B44a - Dutch waterside development, The Hague

Figure B44b - Dutch waterside development, Amsterdam

Figure B44c - Riverside play areas, Kamo River in Kyoto , Japan

Figure B44d - Water meadows and waterside in Cambridge

Rain / Energy Corridor
- Informal recreation
- Footpaths and cycleways
- Energy corridor for wind turbines
- Capacity for attenuation

Home Zone
- Dry access mews parking areas
- Rainwater harvesting
- Gravel swale thresholds

'Causeway Roads'
- Riparian and wet woodland habitats

Room for the River
- Connection with national cycle routes
- Resilient play areas
- Riparian and wet woodland habitats

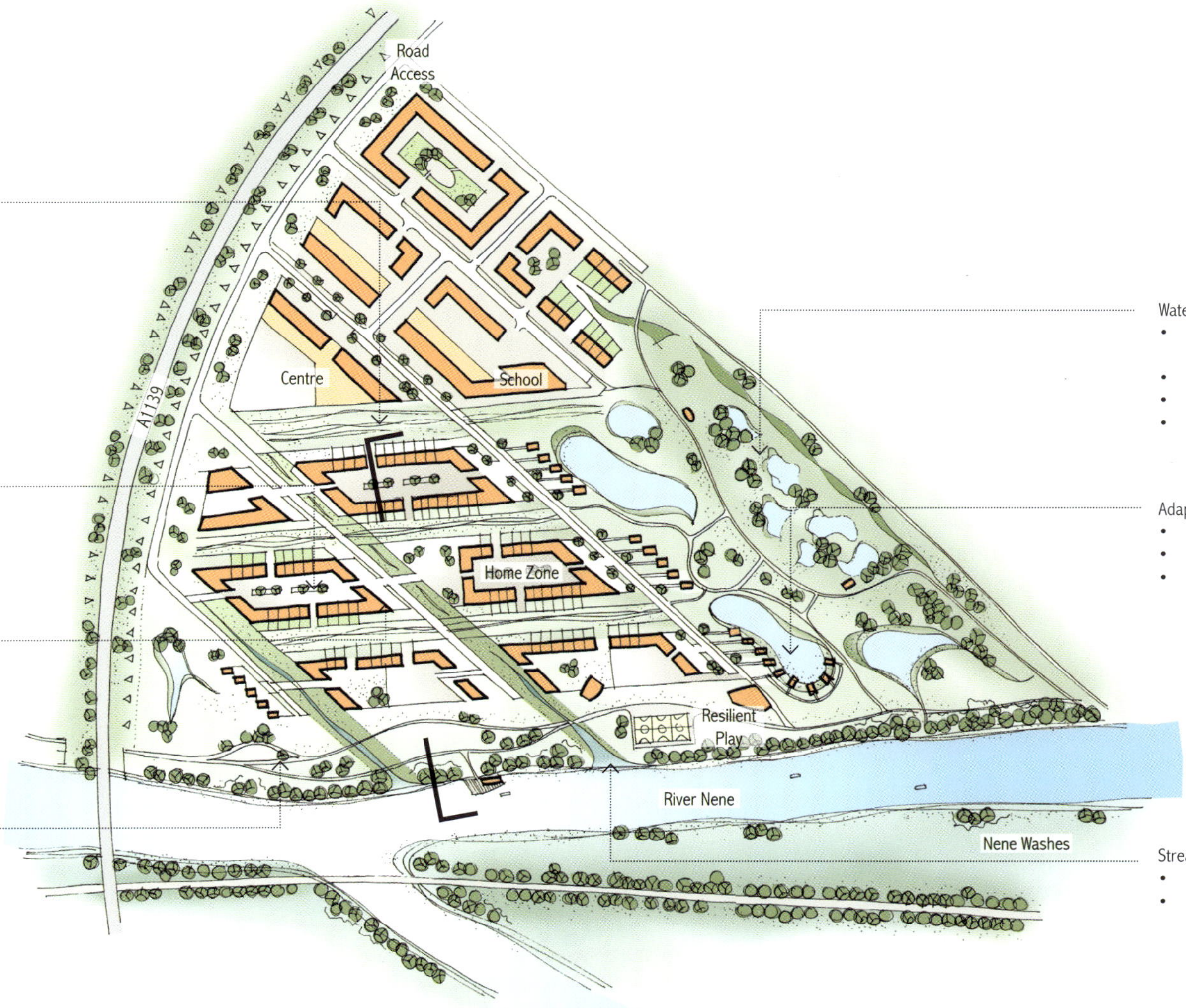

Water meadows
- Publicly accessible nature parkland
- Mixed habitats
- Recreation ponds
- Wildlife habitat

Adaptable buildings
- Amphibious homes
- Stilted office buildings
- Resilient shops

Stream Corridors
- Flood drainage
- Wetland habitats for biodiversity gain

Figure B45 - Site plan

Indicative Sections

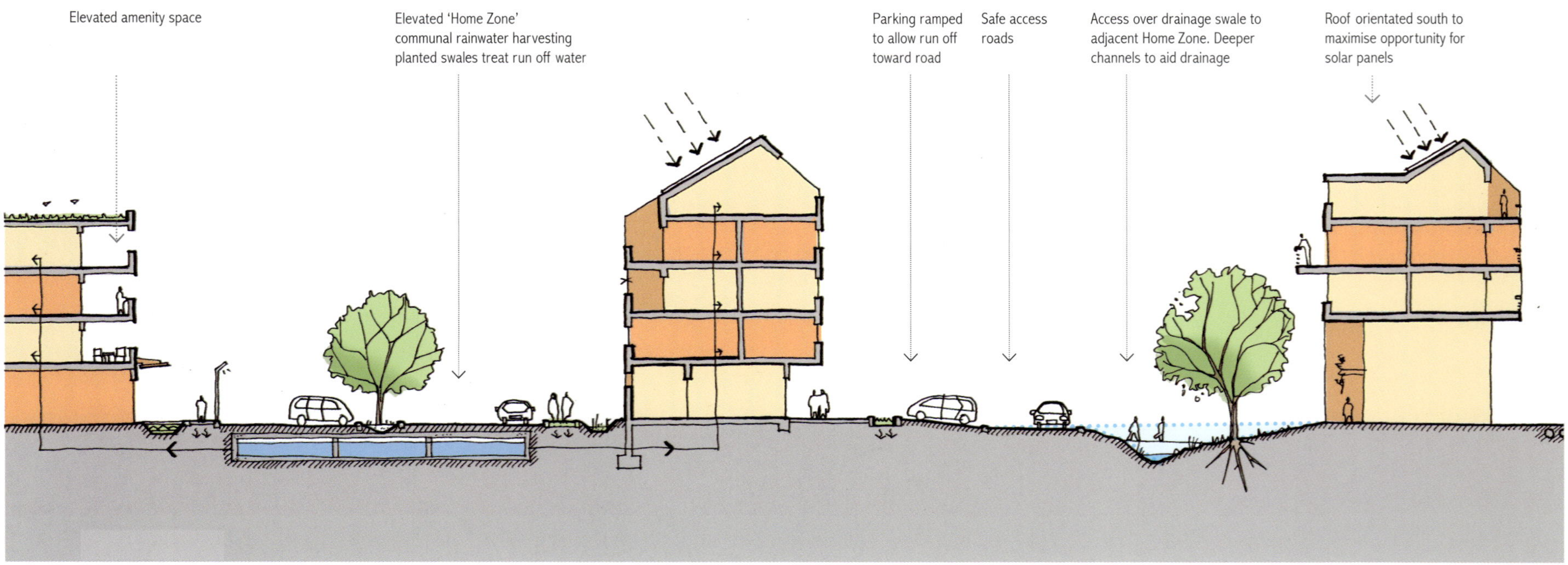

Figure B46a - Section through home zone and stream corridor

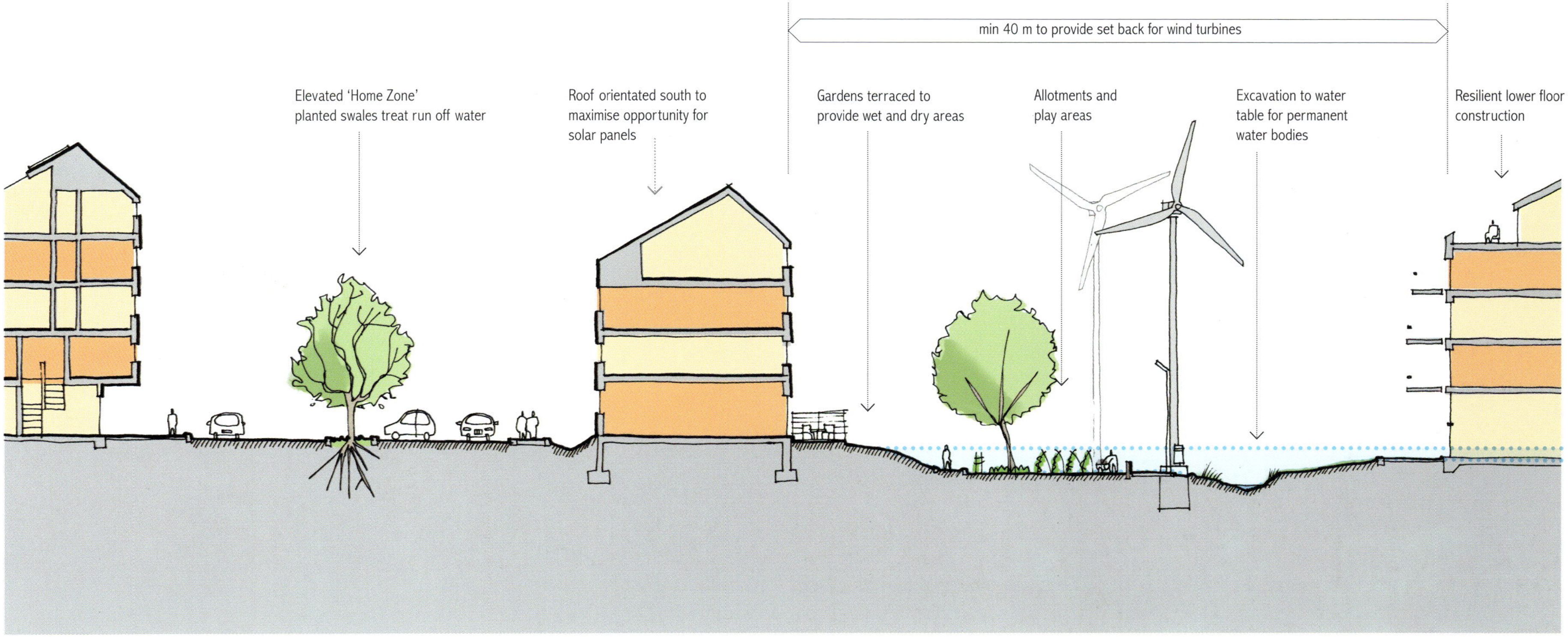

Figure B46b - Section through rain corridor

Indicative Sections

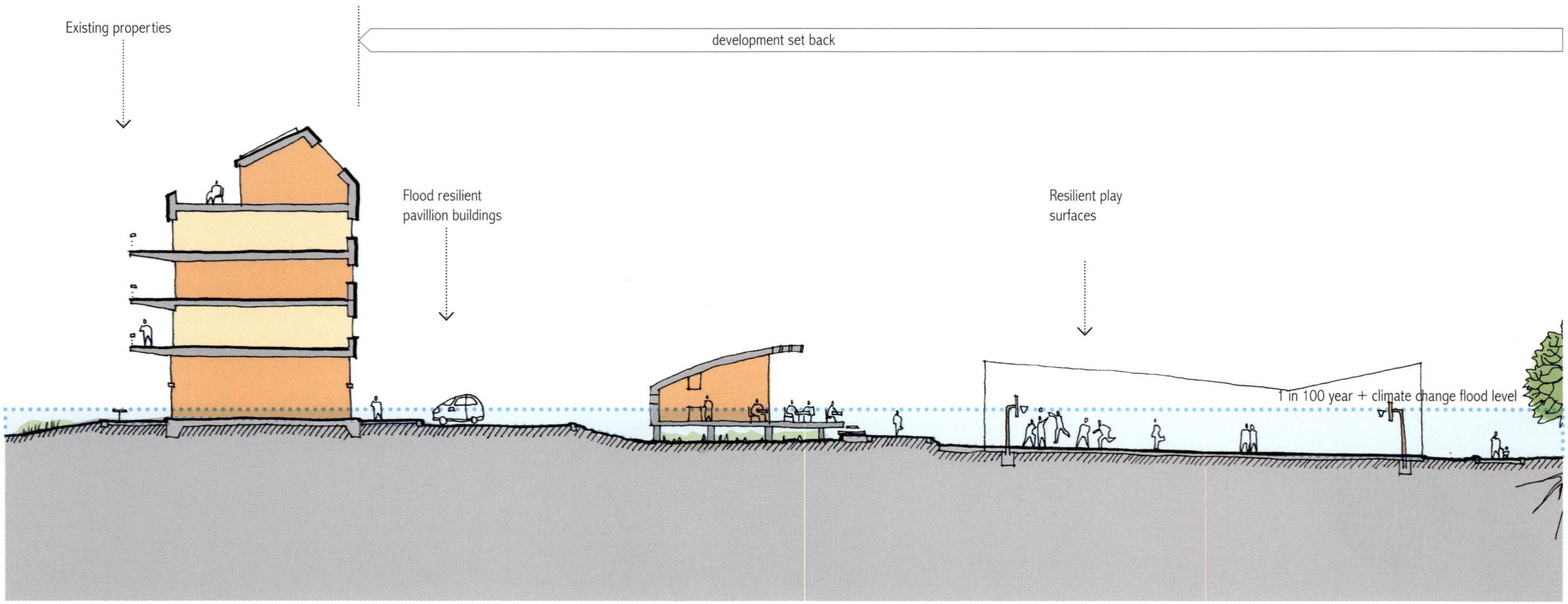

Figure B46c - Section through the River Nene

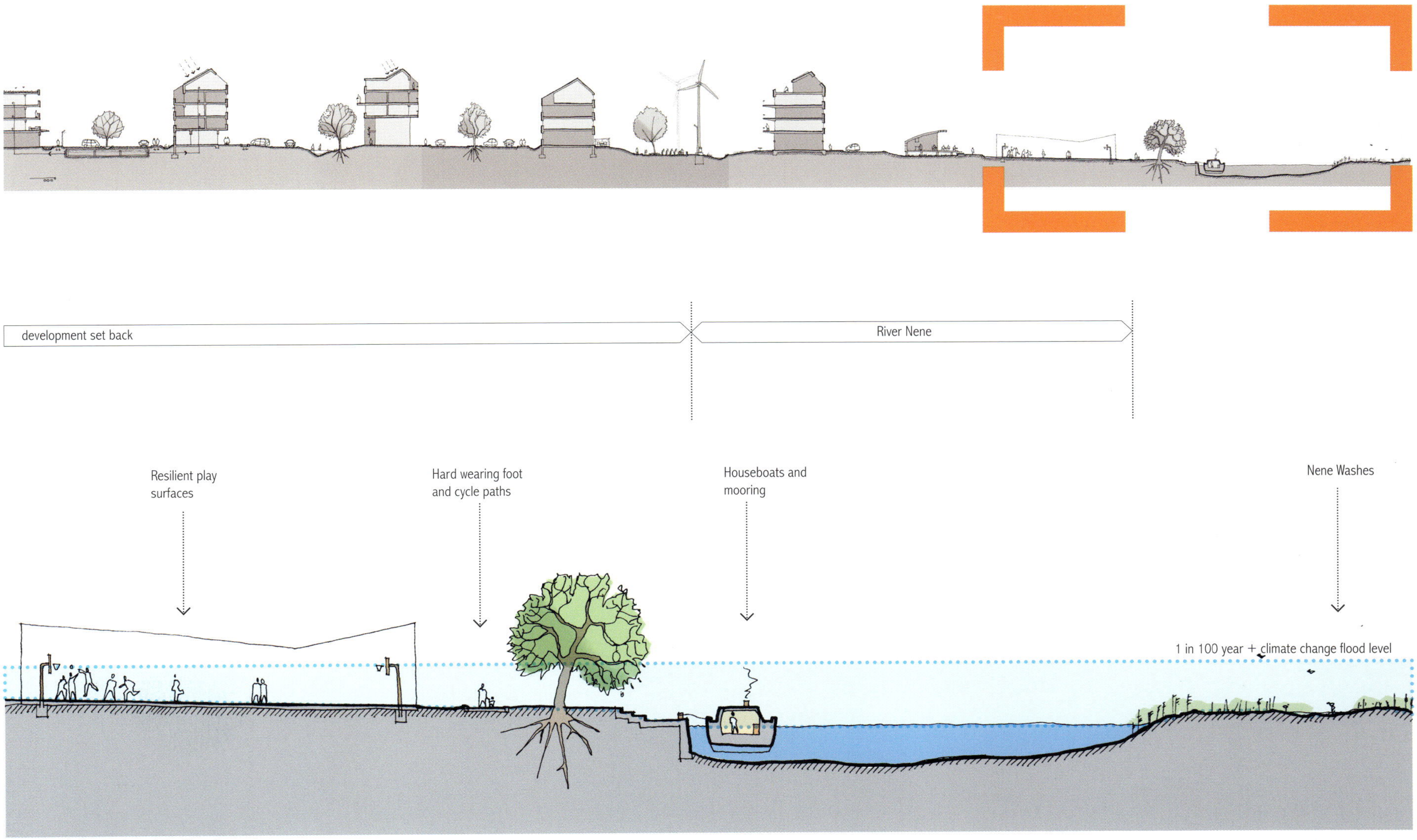

Figure B46d - Section through the River Nene

Figure B47b - During a 1 in 100 year storm the stream corridors channel rain and floodwater off the site to minimise disruption.

Figure B47c - During a 1 in 1000 year storm the floodwater extends over some roads but still allows safe access to homes and property.

Figure B47a (left) - View showing linear parkland drainage corridor and raised access to Home Zones

Review >

The design proposals for all three case studies are assessed collectively in the main report (pages 26-27). A summary of the main issues, particularly with regard to flood risk management, is given below.

Affordability

The total cost of the development for the Riverside sites, built to Zero Carbon, Code for Sustainable Homes Level 6 was calculated at approximately £475,000,000 for options 1 and 2.

Table B5 shows a breakdown of the costs. Figure B49 indicates the proportional cost breakdown.

Option 1 was projected to increase costs by 35% over a traditional development built to current regulations (2008).

Option 2 was projected to increase costs by 32% over a traditional development built to current regulations (2008).

The difference in costs of the two options was a result of the mix of solar PVs and wind turbines. Option 1 which used more smaller wind turbines and more solar PVs was found to be approximately £12m more expensive.

The flood risk management costs were found to be minimal in the base case.* These were found to rise to approximately 1% of the total cost, mostly on the basis of providing SUDS. In addition the cost of resilient building measures resulted in an increase of 11% to the construction cost.

The use of SUDS drainage channels and swales would allow the amount of conventional drainage to be substantially reduced in the LifE options. This was found to reduce the utilities cost. However, the overall energy and utilities cost rose as a result of the renewable energy technologies. The energy and utilities represented 8% of the total base cost and 20% of the total LifE cost, a 234% rise.

The transport and land use costs were found to be cheaper under the LifE options. This was due to the reduction in hard surfaces required, in particular as a result of creating compact Home Zones instead of streets.

A combination of 15 kW and 150 kW turbines as suggested as the ideal solution would result in a cost between options 1 and option 2.

* If the defended option were required to compensate for the loss of floodplain for the future/climate change scenario it is likely that the LifE option costs would be far more comparable if not cheaper.

Site 2, Development Costs

	Base Option Costs	LifE Option 1	LifE Option 2	Percentage change from base (Option 2)
Construction	£250,169,090	£277,978,132	£277,831,342	11.06%
Transport and Land Use	£66,621,953	£62,169,273	£62,067,613	-6.84%
Flood Management	£475,000	£4,809,590	£4,801,770	910.90%
Energy and Utilities	£28,002,000	£105,425,980	£93,465,400	233.78%
Other Code Measures	£9,746,750	£30,497,540	£30,757,390	215.57%
TOTALS	£355,015,000	£480,881,000	£468,924,000	32.09%

Cost plan exclusions:
1) Professional fees
2) Land acquisition costs
3) S106 & S278 costs
4) Service diversions
5) Protection of existing structures
6) VAT
7) Inflation beyond 1Q 2008
8) Costs arising from archaeological investigations
9) Demolition of existing structures on the site
10) Public consultation costs
11) Legal fees
12) Finance charges
13) Sales and marketing costs
14) Compensation costs
15) Amounts payable for continuing maintenance & commuted sums
16) Life cycle & maintenance costs
17) Land remediation costs

Table B5 - Cost summary

Costs

Integrated approach

The cost of developing the site to Code Level 6 and combining ecological flood mitigation in a non-integrated way was assessed to be slightly more expensive than with the LifE integrated approach, see figure B48. The integrated approach was found to save upto £16.4m, 3.4% of the total development cost.

This assessment was made on the basis that without establishing an integrated plan large wind turbines would be untenable, and the area of land required for parking, recreation and SUDS would be greater unless land uses where combined.

The higher density development islands would enable space within the development for wind turbines; and by reducing car provision through creating high density development with short walking distances to car clubs and transport links.

In addition to the financial savings there was very little unused amenity space within the development plans resulting in a high quantity provision for residents.

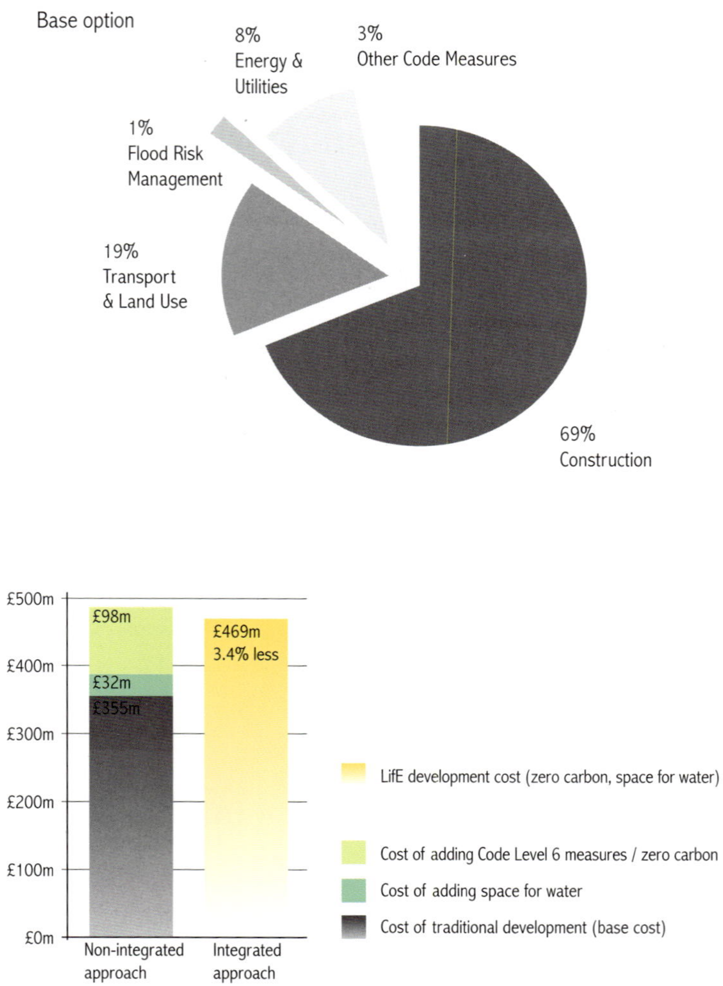

Figure B48 - Cost between integrated and non-integrated approach

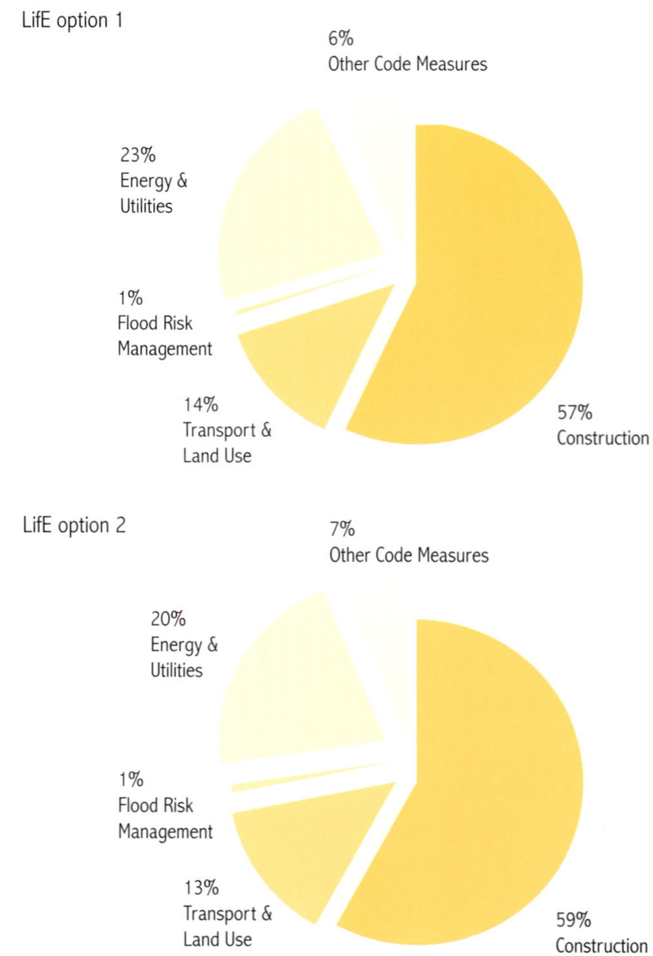

Figure B49 - Cost comparison between base and LifE option

Acceptability

The LifE project proposals were assessed against related planning policy, relevant building regulations, the Code for Sustainable Homes and a number of other guidance documents.

In general the proposals for site 2 were found to be in accordance with policy, particularly planning policy. The integration of amenity with biodiversity, energy generation and water management is well supported in policy.

The site could potentially support large wind turbines (subject to detailed design), however it was felt that local objections would preclude their inclusion; therefore the options focused on the use of smaller wind turbines.* The 150 kw wind turbines would be 45 m high to the top of the rotor blade (equivalent to a 15 storey building) and the 15 kw wind turbines 20 m high (equivalent to a 6/7 storey building). Therefore, these may still meet with local opposition on the basis of visual and noise pollution and the arguable threat to bird life on this recognised flight path of migratory birds.**

Appropriateness

Based on the CMFP modelling information the LifE proposals do not conflict with land use policy guidelines, therefore development would seem appropriate. Planning based on climate change extremes set above national predictions could be considered unduly cautious. However, a precautionary approach, if supported by more detailed modelling, would provide good justification for development of the site to be considered.

The opportunity to reduce housing demand on a site that has a well documented flood history and meets policy would seem appropriate. However, the impacts of climate change must be considered.

* Peterborough already has a number of large wind turbines within its environs, with several more in various stages of being erected. Additional large turbines within the Fengate site could be determined to have an adverse cumulative effect on the City.

** Opinions are divided on whether wind turbines detrimentally affect bird paths or not.

Adaptability

The potential effects of climate change on flood risk on site 2 are three-fold. Increased flood risk from the river, increased risk of flooding from the land and increased risk of tidal flooding resulting from sea level rise and increased storminess.

The potential for change should either preclude development of the site altogether or mean that adaptability of any proposed development be incorporated. The objective of the LifE proposals was to establish a masterplan that would allow the use of the site for development now, but which could allow both buildings and infrastructure to be removed in the future.

There are likely to be many obstacles to achieving land sales and development on the basis that a site may be made floodplain in the future. This approach would favour leasehold sales or short-term licenses to build.

This would allow development to occur in high value riverside sites but provide the mechanism to recover the land in the interest of future flood management. This flexible approach could provide a more adaptable approach to marrying development pressure with flood risk management, than a policy of permission or refusal.

The hierarchical organisation of buildings such that those closest to the source of the risk are more adaptable and more capable of resisting a flood is sensible.

Flood Risk Assessment

1	**Development description and location**	**The River Nene at Peterborough**
1a	What type of development is proposed and where will it be located?	Mixed-use development on the north and south banks of the River Nene in the centre of Peterborough.
1b	What is its vulnerability classification?	Combination of more and less vulnerable
1c	Is the proposed development consistent with the Local Development Documents?	Yes
1d	Please provide evidence that the Sequential Test and where necessary the Exception Test has been applied in the selection of this site for this development type?	N/A. The three study sites were selected during Stage 1 of the project from a number of real sites submitted by local government from around the country. The three sites were chosen on the basis of providing a range of detailed issues to test the 'LifE' strategy and providing a number of flood risk issues that would be most representative of issues likely on other sites.
2	**Definition of the flood hazard**	
2a	What sources of flooding could affect the site? (see annex C PPS25).	Flooding from the River Nene Flooding from land, theoretically but not significant in practice.
2b	For each identified source, describe how flooding would occur, with reference to any historic records wherever these are available.	The flood risk at Peterborough results from the slow build-up and long duration of the flood hydrograph through the upper and middle catchments of the River Nene and overtopping of the river banks. There was a major flood (1% event) on the River Nene in 1998, when Northampton was seriously flooded but Peterborough was relatively little affected.
2c	What are the existing surface water drainage arrangements for the site?	Unknown. Not investigated as part of this study. The scope of work for this study did not involve the development of full FRAs for each of the three sites.
3	**Probability**	
3a	Which flood zone is the site within?	Flood Zones 1, 2 & 3, although little is in Flood Zone 3 and none in the functional floodplain (Zone 3b)
3b	If there is a Strategic Flood Risk Assessment covering this site, what does it show?	Not available at time of the investigation and assessment.
3c	What is the probability of the site flooding, taking account of the contents of the SFRA and of any further site-specific assessment?	Predominantly medium probability – the probability of flooding is unlikely to change as a result of the SFRA. The flood hydrograph will continue to develop slowly and be of long duration (several days), driven by what happens upstream in the catchment and not locally.
3d	What are the existing rates and volumes of run-off generated by the site?	Unknown. Not investigated as part of this study. The scope of work for this study did not involve the development of full FRAs for each of the three sites.
4	**Climate change**	
4a	How is flood risk at the site likely to be affected by climate change?	Increased flows will increase the depth, extent and volume of the river flooding for a particular frequency event, with a consequent change in flood zoning. However, the impact in Peterborough will depend on mitigation measures in the upper and middle catchment.
5	**Detailed development proposals**	
5a	Please provide details of the development layout, referring to the relevant drawings.	*See drawings*
5b	Where appropriate, demonstrate how land uses most sensitive to flood damage have been placed in areas within the site that are at least risk of flooding.	Land use proposed for the sub-areas of the development site is appropriate to the flood risk, with only one small plot to be developed in Flood Zone 3a. Higher density of residential development has been located away from the river in Flood Zone 1.
6	**Flood risk management measures**	
6a	How will the site be protected from flooding, including the potential impacts of climate change, over the development's lifetime?	Flood defences are not proposed but compensatory floodplain storage will be incorporated into the development, and the location and orientation of buildings will facilitate conveyance/drainage towards the river.
7	**Off-site impacts**	
7a	How will you ensure that your proposed development and the measures to protect your site from flooding will not increase flood risk elsewhere?	See above, no defences and floodplain storage will ensure no increase in flood flows upstream or downstream. SUDS systems will provide for 'no detriment', and in all probability improvement, to the surface water flooding situation.
7b	How will you prevent run-off from the completed development causing an impact elsewhere?	See 7a above.
8	**Residual risks**	
8a	What flood-related risks will remain after you have implemented the measures to protect the site from flooding?	See above, the site will not be protected from flooding. Flood risk will increase with climate change and the development planning, location, level and design of buildings, takes this into account.
8b	How, and by whom, will these risks be managed over the lifetime of the development?	Environment Agency = river channel maintenance; Floodplain storage ponds, SUDS systems, Resilient Building Measures etc.

TABLE B6 - Flood Risk Assessment (FRA)

Conclusions

The proposals for the sites at Peterborough showed that the flood risk could be managed with predominantly non-defensive measures. These measures included:

— avoiding risk by locating the highest density development in the areas of lowest probability of flooding, locating buildings on islands of higher land and locating less vulnerable uses in the higher risk areas;

— reducing the extent of flooding by providing rainwater attenuation;

— mitigating the effects of flooding by use of resilient building construction and providing dedicated channels to return floodwater to the river channel after a flood.

These measures provided opportunity for additional recreation space, such as allotments and public parkland, to be created within the development, and renewable energy generation from wind turbines.

The volume and depth of water that would result from a flood at Peterborough were too great to consider flood storage and attenuation measures on site. The conceptual plans for site 2 showed that ecological flood mitigation measures were limited to reducing the effects of a flood on the development. Attenuation of rainwater flows was not thought to reduce the risk during big events.

Spatial planning to locate buildings on the highest parts of the site and creation of clear paths for floodwater to retreat from the site were thought to provide the greatest benefit to reducing risk.

This and the reduction of obstacles to flood flows were considered to be generally applicable to sites located within the middle catchment of rivers.

Safe and predominantly dry access for the lifetime of the development required areas of land to be raised and therefore floodplain compensation to be provided. In areas of greater probability of flooding, compensation volumes are likely to be greater, and would typically need to be provided on land that is upstream of the site.

Resilient construction of buildings was considered important to mitigate the uncertainty of the future flood risk. Amphibious and floating buildings were considered possible for areas closer to the watercourse but would require barriers to flood water debris.

Zero carbon development was made feasible by the use of wind power, to reduce the area of solar panels needed. Heat exchange with the aquifer helped to reduce the total energy demand.

The green channels for flood flows, drainage and rainwater attenuation provided sufficient space for the outdoor play requirements for the residents of the new development.

Figure B50 illustrates the spatial requirements for the sites and the various provisions for amenity space and flood mitigation.

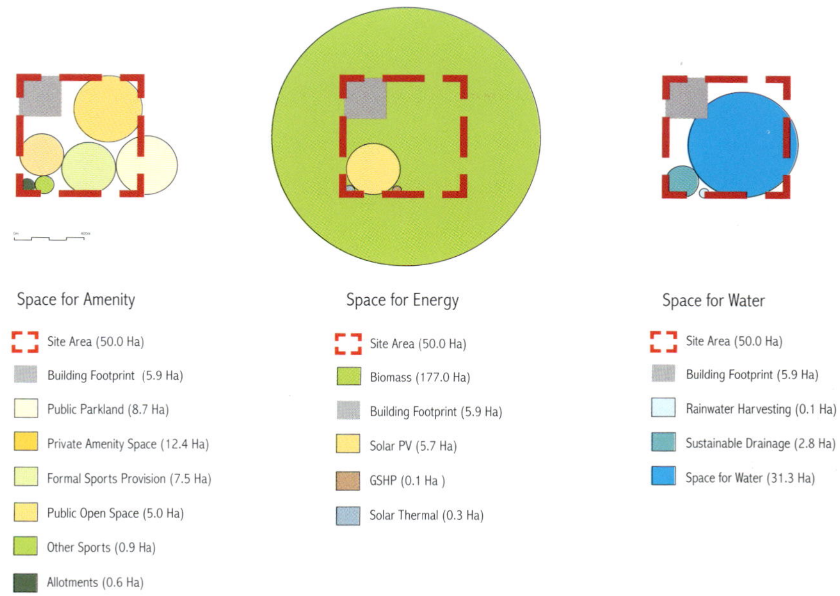

Figure B50 - Space demands for amenity, energy and water on the riverside sites in Peterborough

The middle catchment

The key message for this site was to allow floodwater to flow either through the site or off the site in the event of a flood, to avoid increasing risk up or downstream of the site and reducing the extent of flooding on the site.

'let rivers flow'

APPENDIX C

Site 3, River Arun, Littlehampton

Contents

SITE 3 > The River Arun at Littlehampton

The River Arun at Littlehampton

Site Analysis >

SITE 3 > The River Arun at Littlehampton

Site

Location....................West Sussex
River..........................Arun
Catchment.................Lower Catchment
Area...........................56 hectares developable
Units...........................2200
Density.......................39 units/hectare

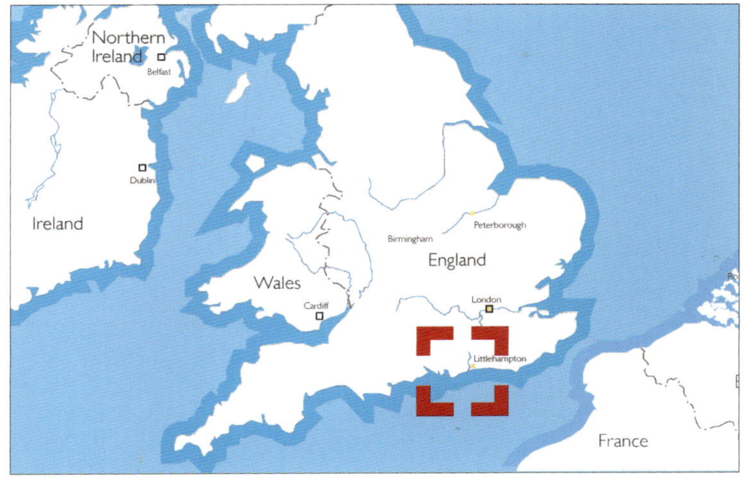

Figure C1 – Location plan

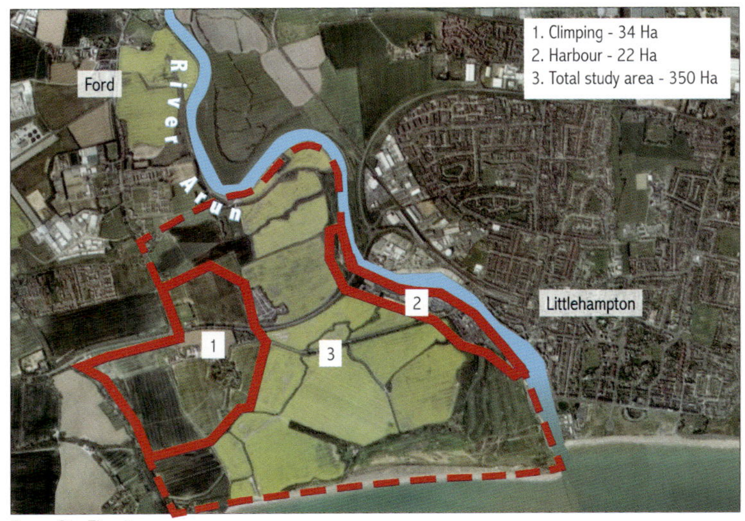

Figure C2 - The site

Understanding the Location

The site is located between the towns of Littlehampton and Bognor Regis, on the south coast of England (see figure C1). The River Arun flows through the site, passing along the western edge of Littlehampton.

The site region (shown dashed in figure C2) is mostly low-grade agricultural land, bisected by the A259 road. A caravan park is located just to the north of the A259. A number of commercial wharfs, a golf course and sewage works are located to the east of the site, on the west bank of the river.

The 'draft South East Plan'[45] recommended 9,300 homes be provided within the Arun District by 2026. For the purposes of this study a requirement to provide 2,000 homes within the site area has been adopted, this number has been based on the expected developer contributions that would be required to provide funding for significant coastal defence improvements that may be necessary over the coming decades. The 'Littlehampton Vision'[46] identified the existing harbour, on the western bank of the river, as a potential regeneration site.

A branch line connects Littlehampton railway station with the main West Coastway Line. A regular service operates to adjacent towns, which in turn have a more regular connection with London Victoria. The site is not well served by buses and the car is the main mode of transport.

The total area of the study site was approximately 350 hectares, much of which is well below the current (2007/8) mean high water spring tide level (MHWS) and would be unsuitable for development of housing. Development is therefore likely to be concentrated in higher density clusters on higher land.

Site Photographs

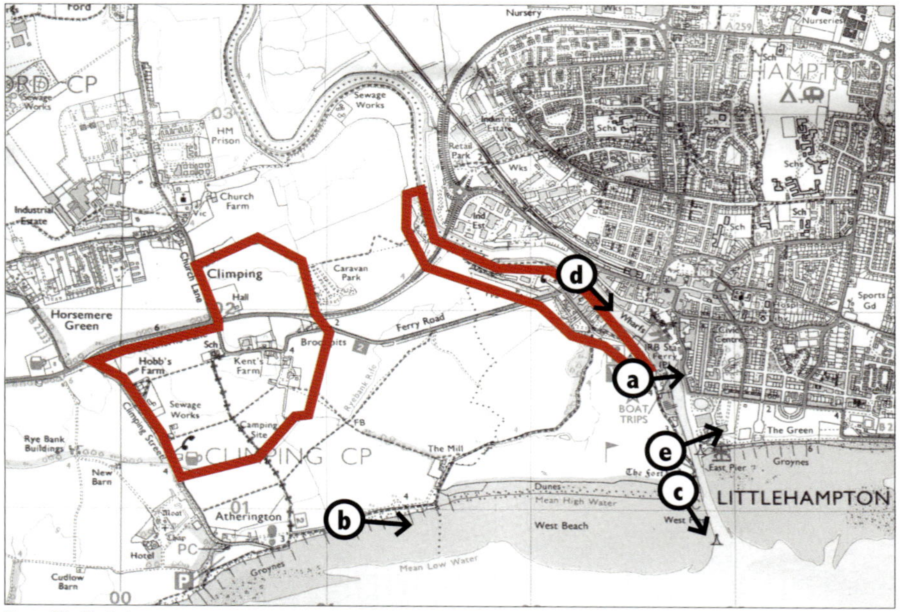

Figure C3 - Photo key

Figure C3a - View towards Littlehampton

Figure C3b - Shingle beach at Climping

Figure C3c - Training wall at mouth of the Arun

Figure C3d - Littlehampton Harbour

Figure C3e - View towards Littlehampton, across the mouth of the River Arun

Littlehampton dates back to Roman times when it was established as a port town, known as Hampton. In the 14th century it became known as Littlehampton to differentiate it from (South)hampton.[42] Littlehampton was predominantly a maritime town until the railway was opened in 1863 when it began to attract seaside tourists. It is now an established seaside resort town (see figures C3a to C3e)

Historical maps (shown in figure C4) show that there have been a number of changes to the River Arun over the last few centuries. A map from 1698 (figure C4c) shows that the river flowed into the sea via a series of channels cut through the beach. The 1879 first edition Ordnance Survey map (figure C4d) shows the river training walls in their current alignment, constricting the mouth of the river. Dunes were indicated along the west bank of the river and an area of reclaimed land is shown inshore of the dunes. Since 1879 the alignment of the river has remained unchanged, and the town of Littlehampton has expanded.

Medieval earthworks have been found to the east of St Mary's Church, Climping and Medieval and Romano-British finds indicating former house sites have been found further north.

A 19th century fort to the west of the river mouth, built as part of the south coast defences to protect the entrance to the River Arun in 1854, is now overgrown and in poor condition.

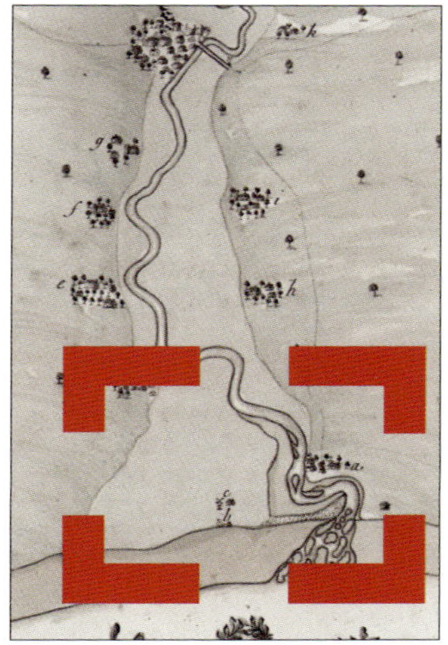

Figure C4c - 1698 South Coast Harbours Report, Edward Dummers and Thomas Wiltshaw

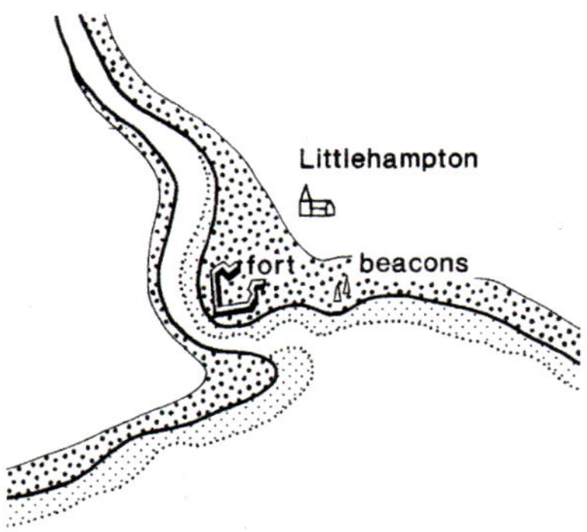

Figure C4a - 1587 Armada map

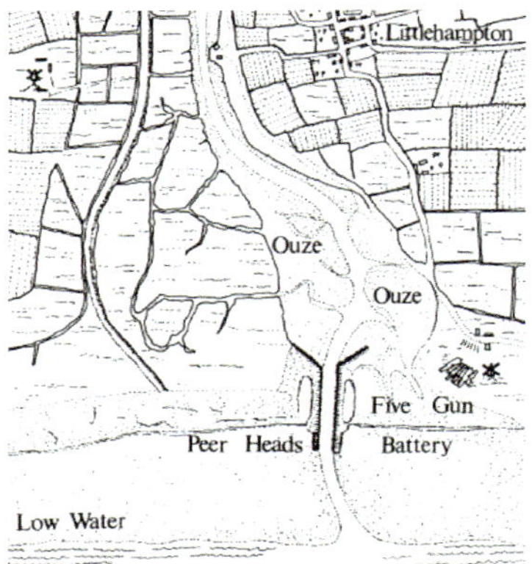

Figure C4b - From an early plan of Arundel Haven about 1736.

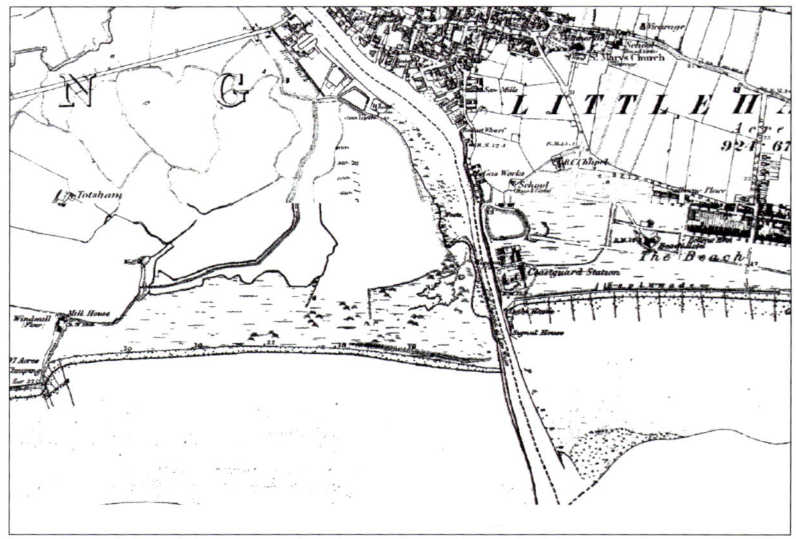

Figure C4d - 1879 (1st Edition Ordnance Survey)

The River Arun

The River Arun flows from its source in the St Leonard's Forest area to its mouth with the English Channel at Littlehampton, dissecting the South Downs on route.

The site is located within the lower catchment of the river (figure C8). The lower Arun lies within a flat river valley cut between the South Downs, shown as the area of light green in figure C7. The river has been heavily engineered, such that the former estuarine floodplain has been almost entirely reclaimed for agricultural use (figure C5). Figure C9 shows the land levels above Ordnance Datum. The mean high water springs level (MHWS) is 2.9m above Ordnance Datum. Parts of Littlehampton and Arundel have been constructed below this level.

The mouth of the river has been narrowed from 1000 m, indicated in historic maps, to its current extent of 40 m. A training wall extends some 340 m offshore to maintain a navigable entrance channel. The river is used for recreation and industry (see figure C6).

Fluvial flows are typically sluggish in summer but the river is prone to flash flooding in winter. The river is tidal up to Pulborough, some 18 km inland, and during dry periods the majority of the flow in the lower valley is tidal water.

The mean tidal flow at Littlehampton Harbour mouth has been measured as 135 m^3/s with tidal current velocities of up to 2.5 m/s (SCOPAC, 2004). The estuary is ebb dominant (flow towards the sea) and tidal scour has helped to maintain the inlet and the depth of clearance over the harbour bar. Although the inner shoals are dredged regularly, the material removed from the mouth is replaced with new sediment in relatively short time periods.

Figure C5 - Low-lying farmland to the north of the coast

Figure C6 - The dredging works to the east bank (right) of the River Arun

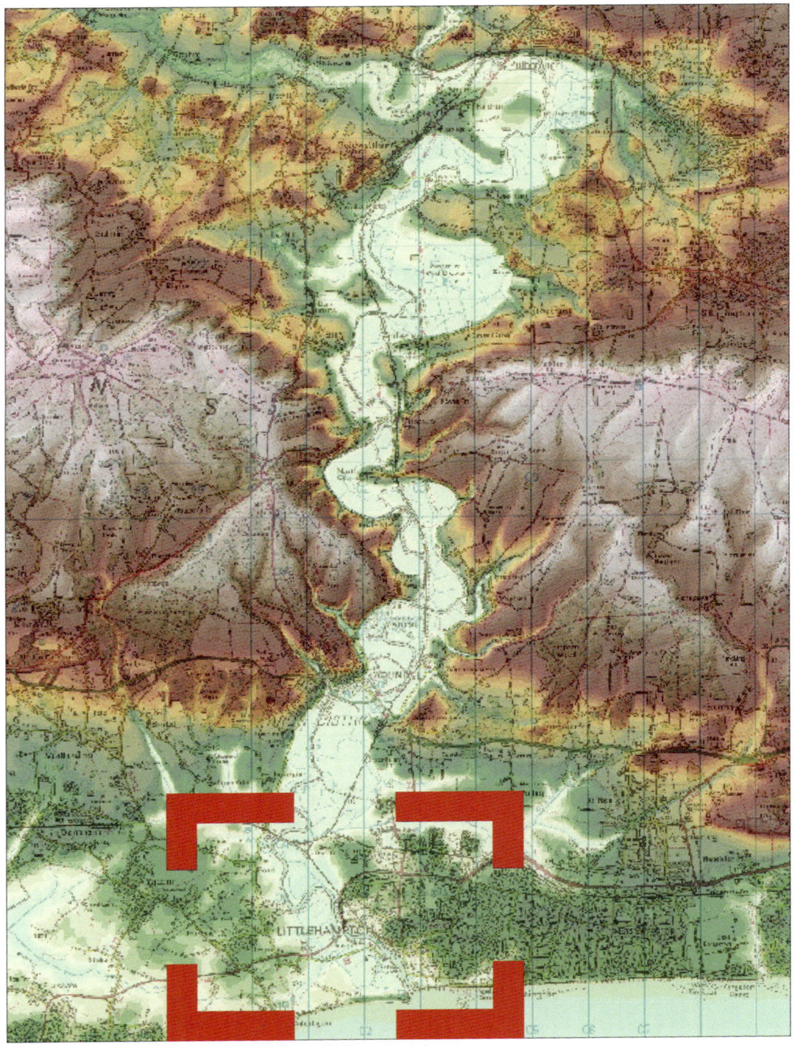

Figure C7 - Topographical map, showing the south downs in pink

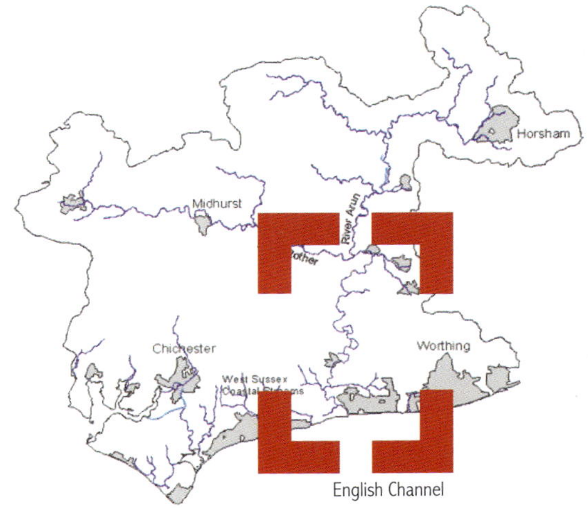

Figure C8 - River catchment system

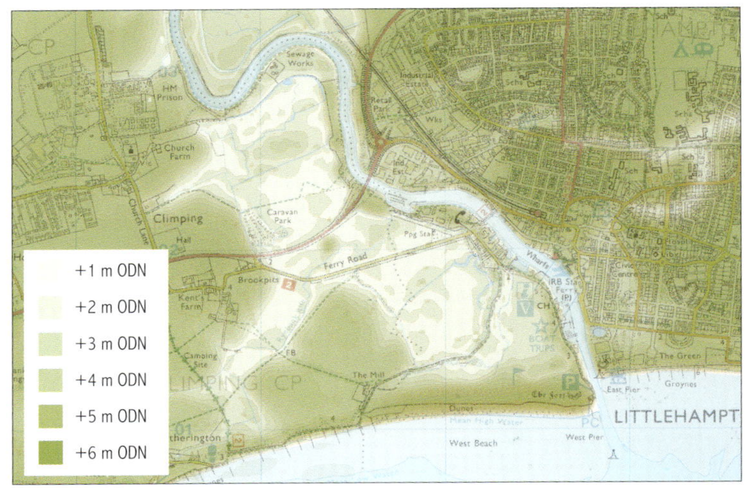

Figure C9 - Contour map

Landscape Characteristics

The lower Arun is identified as a'strategic gap' site, established in the West Sussex Structure Plan to prevent coalescence of separate communities and to protect the wider landscape and environment from detrimental development.

Arun District Local Plan sets out aims to maintain and enhance these gaps by limiting development, promoting the long-term enhancement of the landscape, amenity and conservation value, and improving public access.

The Arun valley is a strongly pastoral valley floor and floodplain, with extensive wet meadows and grazing marsh in the south. The low-lying land is intensively farmed arable land with predominantly large field sizes, enclosed by hedgerows and tree belts. The landscape character is indicated in figures C12 and C13.

To the north of the site the low land provides dramatic views across the river valley to Arundel in the north (see figure C10). To the south of the site the sound of the waves, wind and birds become more dominant, though the views of the sea are obscured by the shingle ridge (figure C11).

Areas of the site are designated as Coastal and Floodplain Grazing Marsh. These are typically characterised by ditches, used to drain the land, which are especially rich in plants and invertebrates. Floodplain grazing marsh is typically important for breeding waders, such as snipe, lapwing and curlew; and for internationally important populations of wintering wildfowl, including Bewick and Whooper swans; though these were not identified within the boundary of the site.

A network of designated public rights of way, shown in figure C13, and cycle routes crosses the land. These routes connect Littlehampton with local communities such as Climping and Atherington.

Figure C10 – View of Arundel from the River Arun

Figure C11 – Climping Beach

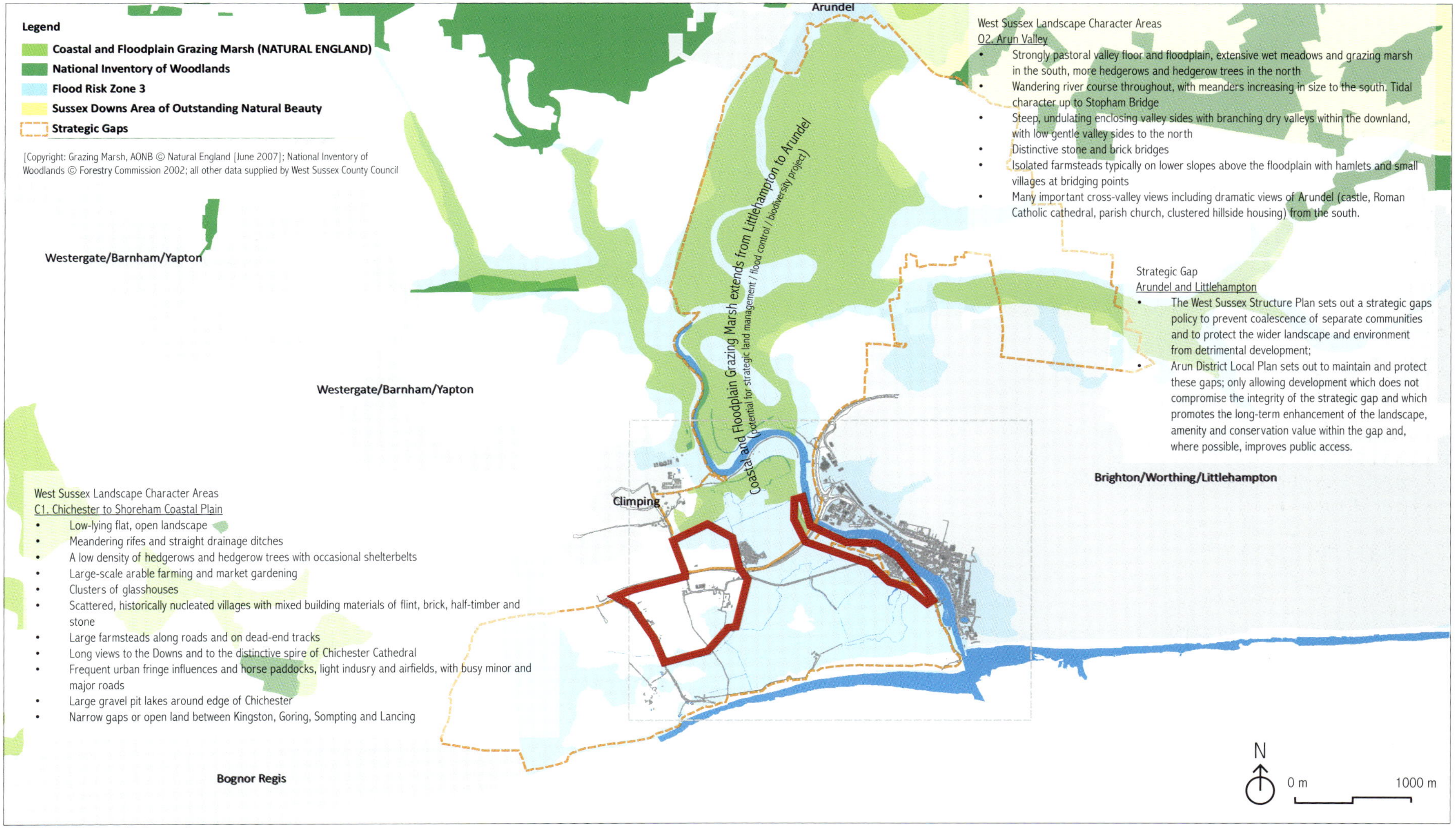

Figure C12 – Regional landscape character

Landscape Characteristics

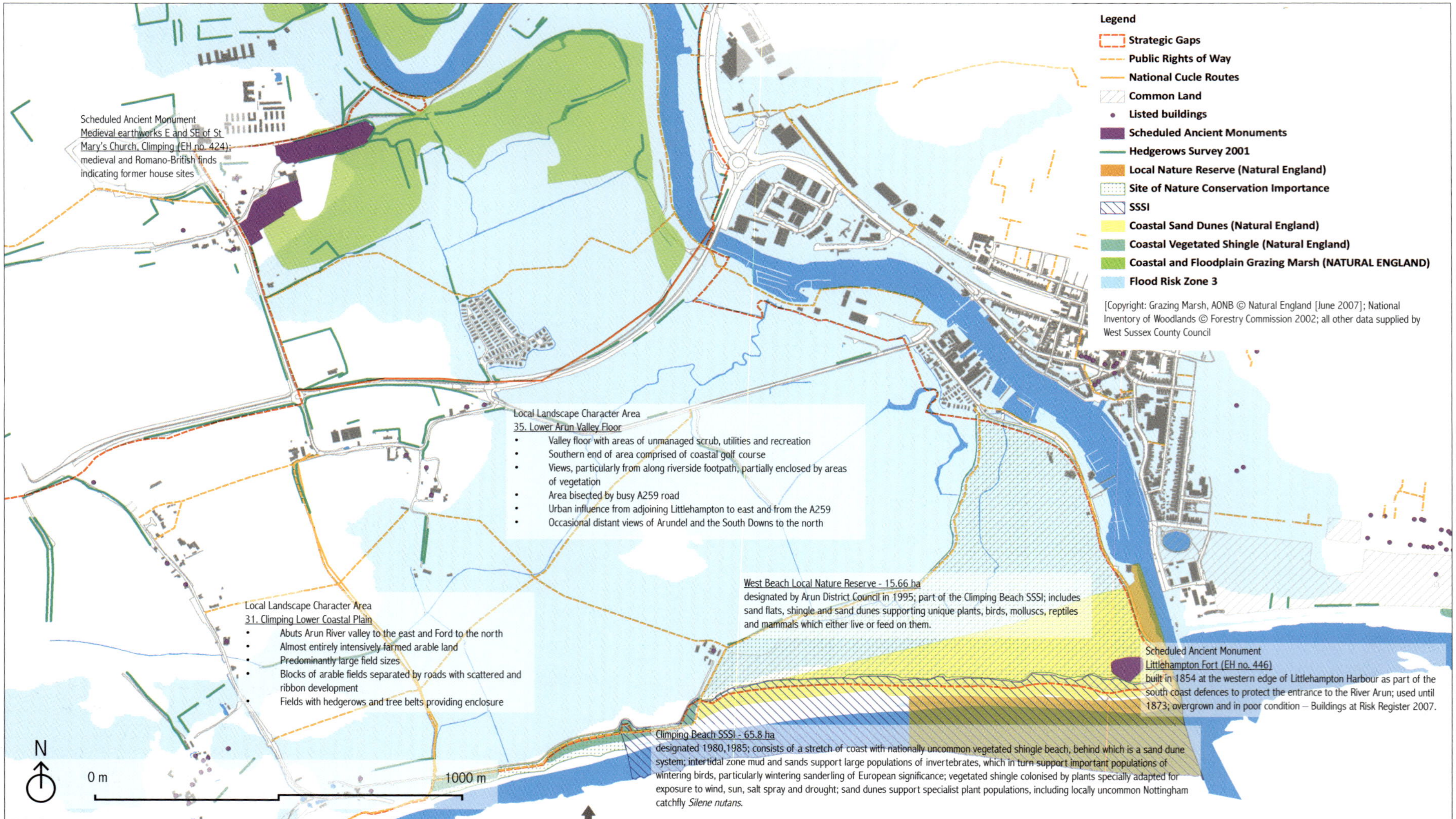

Figure C13 – Local landscape character

Flood risk to the site is dominated by coastal conditions, although fluvial flows could also be significant during a severe event.

The Arun and Western Streams Catchment Flood Management Plan (CFMP) covers the entire lower tidal floodplain including Arundel and Littlehampton where the fluvial flood risk is dominated by the tide.

Surface water drainage is also an issue particularly where prolonged periods of rainfall coincide with high river levels, and surface water drains are tide-locked.

There are currently believed to be 1051 residential and 198 commercial properties at risk from a 1% (1 in 100 year) flood in the lower Arun valley, with likely damages estimated at £165m. With predicted climate change increasing both sea level and river flow, this is expected to double over the next 100 years to damages in excess of £400m.[25]

Extent

The Environment Agency flood risk map (figure C14) shows that the site is predominantly within the 0.5% (200 year) probability flood risk area. The 0.1% probability (1000 year) flood risk area is only shown to be slightly more extensive. The digital flood mapping supplied by the Environment Agency, shown in figure C15, appears to show similar extents. However, areas of Littlehampton appear to be at greater risk of flooding.

Existing Defences

The river mouth is constrained by the existing training walls, which help to maintain a navigable channel. They also reduce the volume of water entering the river channel. These are expected to come to the end of their functional lives in 2017*. Without these walls the the mouth of the river would eventually return to a more natural form, similar to that shown in the map of 1698, with a likely increase to the risk of flooding upstream.

The river defences include a mixture of walls and embankments of various materials in varying conditions and are at risk from overtopping in places. In some places there are no formal defence structures. The line of the defences is complex and heavily influenced by the river and hinterland use. Localised flooding has occurred in the area of the Yacht Club, although the associated damages have been small.[47]

Around the A259 bridge there is low-lying agricultural land behind the defences, which is vulnerable to flooding. It is understood that a section of the west bank of the River Arun was historically used as a refuse tip. This may represent a constraint to implementing managed realignment in this area in the future and it may be necessary to continue to provide flood and coastal erosion risk management measures in order to prevent the release of contaminants into the estuary.

* Residual life estimated at 15 years.[47]

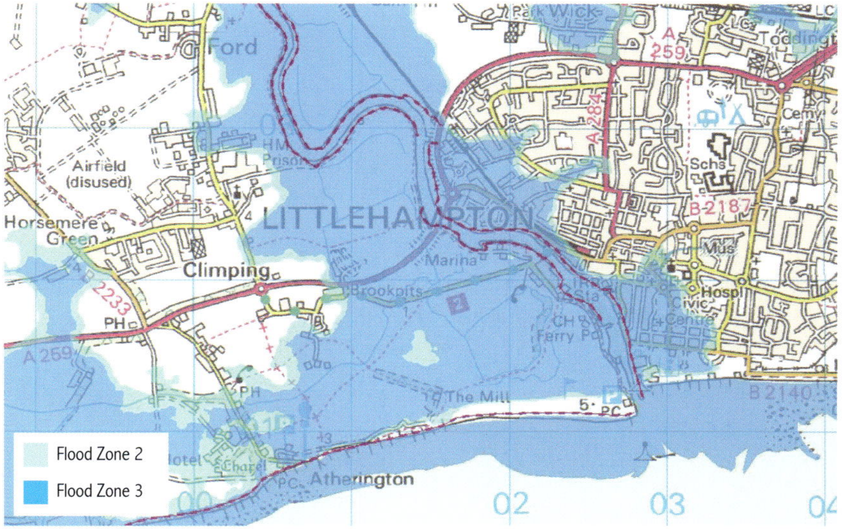

Figure C14 - Environment Agency flood risk map

Flood Zone 1 (Low Probability)
Areas with less than 1 in 1000 year (0.1%) annual probability of flooding.

Flood Zone 2 (Medium Probability)
Areas with between 1 in 1000 year (0.1%) and 1 in 100* year (1%) probability of flooding.
* 1 in 200 year (0.5%) probability of flooding from the sea

Flood Zone 3a (High Probability)
Areas with between 1 in 100* year (1%) and 1 in 20 year (5%) probability of flooding.
* 1 in 200 year (0.5%) probability of flooding from the sea

Flood Zone 3b** (Functional Floodplain)
Areas with greater than 1 in 20 year (5%) probability of flooding.
** This is not shown on the Environment Agency website.

Flood Risk

Duration

Tidal river flooding is likely to be of short duration with tides controlling the rise and fall of river levels. Therefore flooding is likely to retreat with the tide, lasting for a time equivalent to the change from high to low tide, 6.2 hours. However, local conditions will affect the ability for the floodwater to drain away and areas of floodwater could remain on site for days. Additionally, repeat high tides could prevent floodwater draining away, prolonging the period of inundation.

Risk

Flood risk is described in terms of the sources of flooding, the pathway or route by which the flood would occur and the receptor such as homes or farmland; the risk to the site is set out in Table C1.

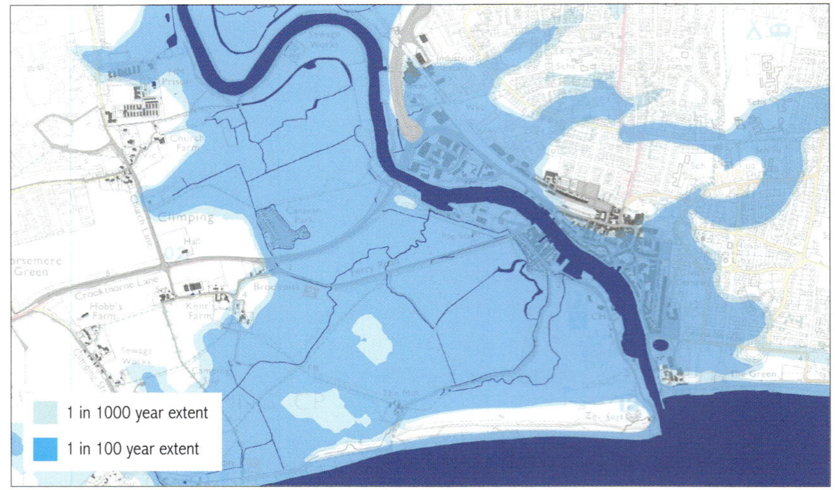

Figure C15 - Flood map, current

Site 3, Flood Risk		
Sources of Flooding	Pathway	Receptor
Flooding from the Sea	Overtopping of sea defences	Farmland
Flooding from Rivers	Failure of river walls	Homes and businesses
Flooding from Land	Overtopping of river defences	Sewage treatment works
Flooding from Groundwater	Surface water run off behind defences	Caravan park
Flooding from Sewers	High tides block drain flows resulting in manhole overflows	Industry & boat yards
	Ground surge or passage under defences	Marina

Table C1 - Flood risk

Climate Change

The southeast is expected to become drier in the future with decreased annual and summer precipitation but potentially more intense rainstorms.

Sea Level Rise

The low-lying nature of the south coast of England means that sea level rise will have a significant impact on the entire region. Wave heights, tide levels and flood flows are expected to increase, resulting in higher flood levels for a particular event. Areas currently not at risk of flooding may become subject to flooding, whilst existing flood areas will be inundated to greater depths.

Figure C16 shows the extent of land that would be at risk of flooding during a 0.5% and 0.1% flood event, with 1045 mm (1 m) sea level rise.

Figure C17 shows the extent of land that would be at risk of flooding during a 0.5% and 0.1% flood event, with 2 m sea level rise. This shows the entire region with the exception of a few high points within Littlehampton to be at risk of flooding.

These figures illustrate the sensitivity of the site to sea level rise.

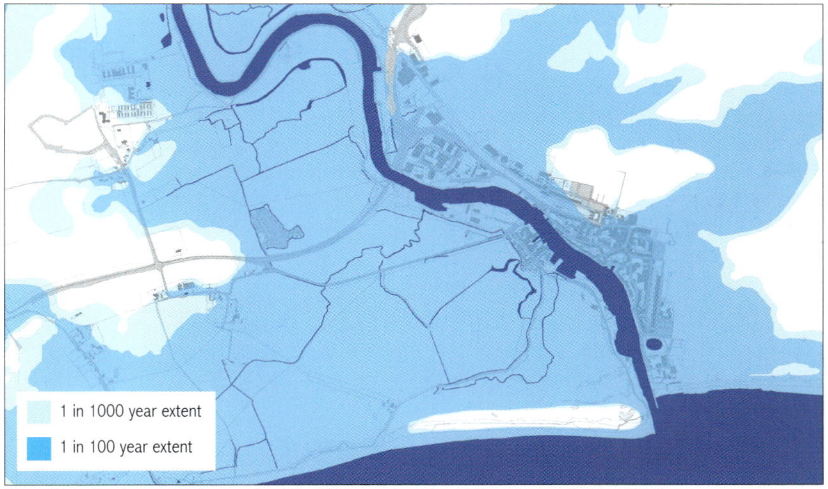

Figure C16 - Flood map, with 1 m sea level rise (100 years based on Defra guidance, October 2006)

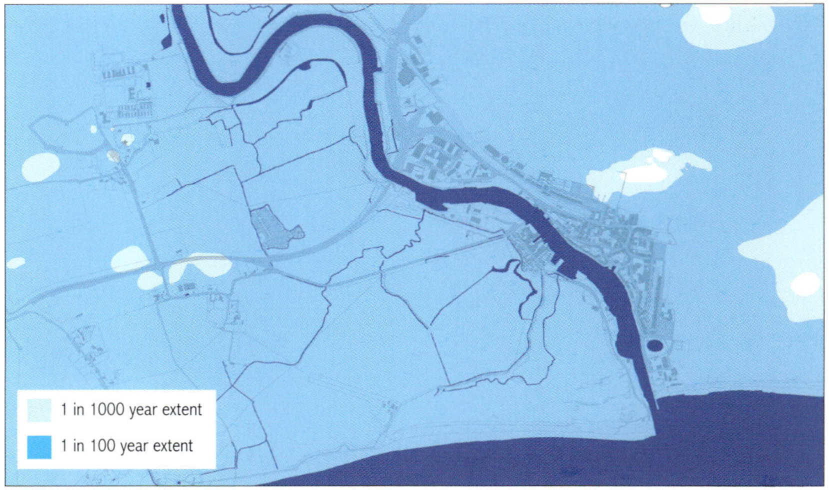

Figure C17 - Flood map, with 2 m sea level rise

Effects of Climate Change

Increased Peak Rainfall

Flood risk as identified in the Catchment Flood Management Plan for the Arun and Western Streams will increase markedly this century as a result of sea level rise. Increased rainfall will increase the risk of flash flooding and pluvial flooding.

Increased River Flows

The cost of damages that would result from a 1% annual probability fluvial flood event in the lower Arun in 2106 have been estimated in the order of £400m* according to the draft Catchment Flood Management Plan.[25]

Increased probability, frequency and depth of flooding will make the provision of safe access and egress from proposed developments during extreme storm events more difficult and expensive.

* It is not clear how this figure was determined and if this assumes that property would be written off as a result of flooding. It is assumed that this scenario would coincide with a high tide.

Energy Sources

The available natural resources that would be suitable for energy generation at Littlehampton are wind, tidal and solar. It is possible that a local aquifer could be used, particularly for low carbon cooling.

The average wind speed is indicated as 6.2 m/s at 25 m above ground level and 6.6 m/s at 45 m above ground level. This suggests that commercial 2 MW wind turbines may be feasible on site.

The land area is also large enough to consider growing biomass on site.

Therefore the following energy technologies were identified as viable on this site:

- Wind
- Tidal (tidal stream or tidal range technologies, such as turbines)
- Solar (Photovoltaic and Solar Hot Water)
- Biomass
- Ground Source Heat Pumps
- Aquifer Thermal Energy Storage (ATES)

Rainfall

The total site area for the main site is 3,541,000 m^2 and the average annual rainfall is approximately 600 mm per year*. Therefore the total volume of rain falling on the site would be approximately 2,124,637 m^3 per year.

The 220,000 m^2 harbour site, on the west bank of the Arun would receive an average annual volume of rain of approximately 132,000 m^3.

* Average annual rainfall varies across the region from 800 to 1,000 mm in the west, to 550 to 650 mm in parts of Kent. Source: Environment Agency

Formulating the Design Brief >

Development Mix

The accommodation schedule, shown in Table C2, was determined from past project experience and research into local demand. Using this schedule, energy demand, water demand and amenity standards could be determined.

A conceptual mixed-use development of 2180 homes was proposed across two sites. A further 27,000 m² of non-domestic buildings were also proposed.

These two sites are Climping/Ford to the west of the site and Littlehampton Westbank Harbour site to the east. These sites could potentially provide over 1000 homes each with development densities of 32 units per hectare and 56 units per hectare respectively.

Amenity Requirement

Based on the accommodation schedule the number of inhabitants was estimated to be approximately 5500.

An assessment of applicable national and local planning policy guidance (based on a combination of NPFA's Six Acre Standard and Arun District Council Adopted Plan 2000) was undertaken to determine minimum standards of open space required for the development.

Figure C19 illustrates the requirements for open play areas for approximately 5500 people, which were:

- Sports and play provision* – 8.77 Ha
- Outdoor (public) open space – 4.38 Ha
- Private amenity space** – 8.42 Ha
- Total required outdoor space – 21.57 Ha

The total requirement for the Harbour site was approximately 10 Ha (based on 2500 people).

* Including formal pitches and other areas of play

** Assumption based on site 1 and 2 requirements

Site 3, Accommodation Schedule

Residential	GIFA (in m²)	No. of Homes	Total Area (inc circulation)	Parking
1 bed studio	37	0	0	0
1 bed flats	50	540	31,050	702
2 bed flats	70	760	61,180	988
3 bed flats	90	220	22,770	440
3 bed houses	100	220	22,000	440
4 bed houses	120	220	26,400	440
5 bed houses	140	110	15,400	220
6 bed houses	160	110	17,600	220
Residential GIFA (in m²)		2,180	196,400	3,450
Commercial			GIFA (in m²)	Parking
Office Area			7,256	363
Retail Area			14,512	726
Industrial Area			1,814	91
Educational Area (inc nurseries, creche..)			1,814	91
Health and Community Facilities			1,814	91
Commercial GIFA (in m²)			27,120	1,361
TOTAL GIFA			223,610	4,811

Table C2 - Accommodation schedule

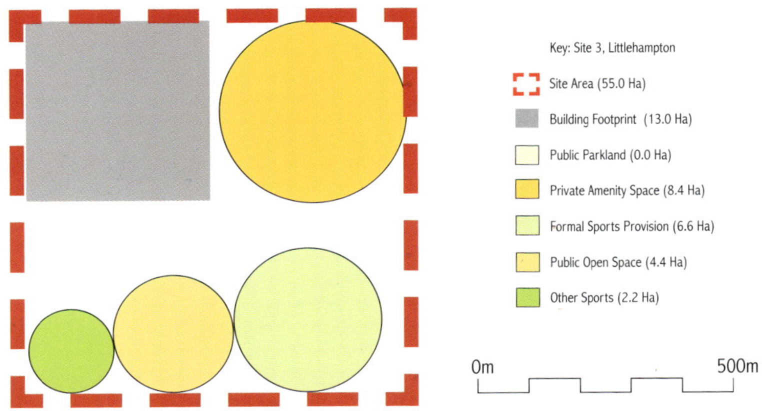

Figure C19 - Spatial requirements for open play

Resources

Energy Demand

Using a range of typical energy consumption figures for different building types (details included in appendix D) the accommodation schedule was used to calculate the total energy demand for the both sites, indicated in Table C3 and Figure C20.

Water Demand

The total water demand was calculated as 662,600 litres per day, 235,200 m^3 per year (based on 110 l/p/day for residential, which incorporates dual flush WCs and low flow taps + 11 l/p/yr for commercial).

This water demand is approximately 11% of the total rainfall across the entire 350 Ha site. More importantly the demand is approximately 70% of the rainfall on the developable site areas.

Site 3, Energy Demand

Energy Demand (MWh/yr)	Space heating	Space cooling	Hot water	Electricity (excl cooling)	Total (MWh/yr)	Percentage
Total Residential Demand	2,720	0	6,015	7,675	16,410	74.6%
Total Commercial Demand	660	1,325	315	3,280	5,580	25.4%
Total Energy Demand	3,380	1,325	6,330	10,955	21,990	100.0%

Table C3- Energy demand

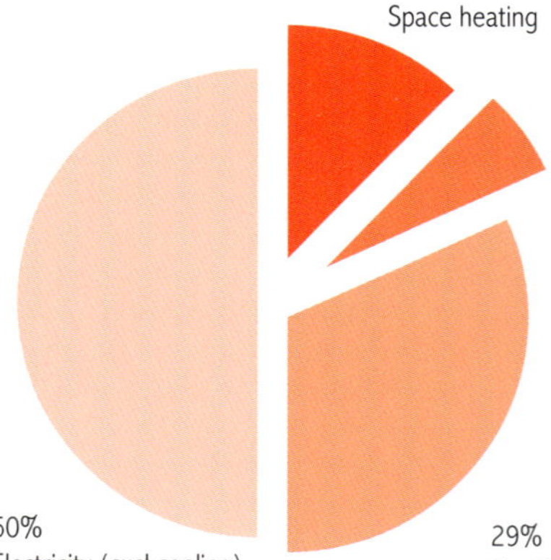

Figure C20 - Energy demand mix

SUDS Requirement

The design standard for surface water drainage is 1 in 30 years, ie that the capacity of the drains can cope with a severe rainstorm, only likely once every 30 years. However, the design standard for flooding is 1 in 100 years. During a 1 in 100 year rainstorm drains are likely to be overwhelmed.

Current EA policy, is to aim for 'greenfield' equivalent infiltration and run off whenever possible.

The design standard adopted by the LifE project is to provide surface water drainage that can cope with a 1 in 100 year event plus a 30% increase in peak rainfall to account for climate change to 2115.

The geology is thought to be sand and gravel over a layer of chalk, which would have a high permeability, however, the groundwater table is likely to be very high, particularly as areas of the site are within the normal tidal range. This would impede the ability for the water to soak away in low-lying areas.

The Harbour/Westbank site is likely to contain areas with contamination and due to the surrounding water bodies could be tide-locked during a storm event, therefore infiltration will not be an option and a sealed system would be required. This system could also be required to store the water above the groundwater table to prevent increasing the risk of groundwater flooding.

The Climping/South Ford site is less likely to be contaminated and is located approximately 2 m above the flood risk level. Therefore it is probable that a permeable, infiltration system would be suitable.

A precautionary assessment for the volume of attenuation would be 300 m^3 rainfall per hectare.* This would equate to a total attenuation volume of 10,200 m^3 for the climping site and 6,600 m^3 for the Harbour/Westbank Site.

The total optimum water requirements are shown in figure C21.

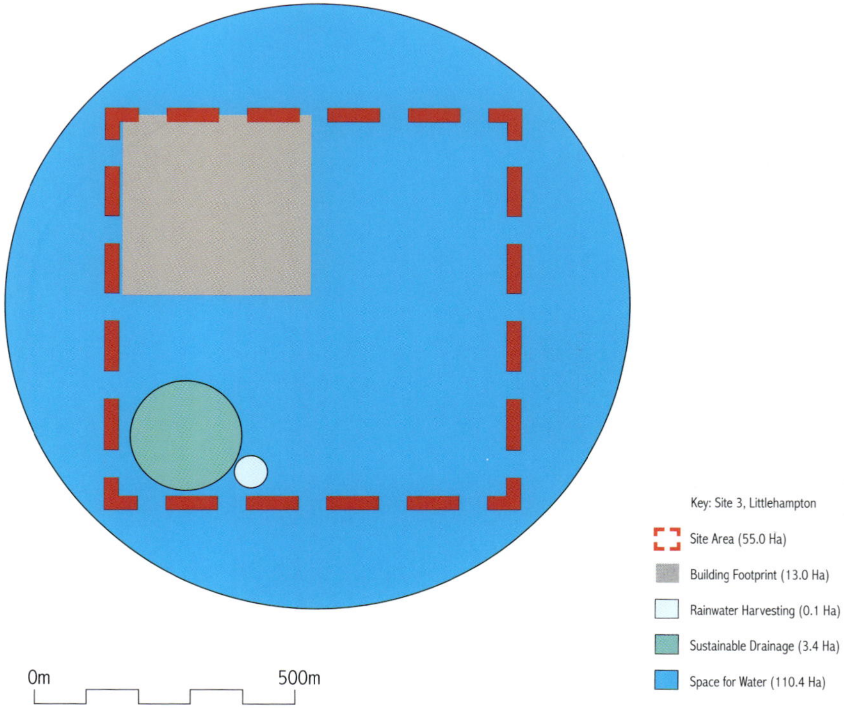

Figure C21- Spatial requirements for rainwater, attenuation, harvesting etc..

* 300 m^3 per hectare equates to 8 l/s/ha for a 10 hour rainstorm or 5 l/s/ha for a 16 hour rainstorm.

Design Objectives >

Design Objectives

The objective was to establish a strategic plan for the lower Arun that would improve regional flood risk management and enable redevelopment of the harbour site in particular.

Adaptable planning, working with natural coastal processes, would seek to maximise benefits to the local community of wetland regeneration and other flood protection measures to offset loss of land.

Design Objectives

The large overall site area and variety of renewable energy sources allow a whole range of options to be considered, particularly for energy and flood risk management. At the same time the severity of the flood risk and implications of climate change require a very careful and consolidated approach to development.

Main Objective

Create a highly diverse landscape to provide multiple benefits to new and existing residents of Littlehampton through flood protection, amenity and habitat generation, tourism and economic sustainability.

Landscape Approach

Large-scale restoration of the lower Arun floodplain would provide a number of opportunities for improvements to the landscape, including improved access, wetland habitat, tourism and recreation and food and energy crops.

Energy and Water Approach

Use the wide variety of resources available to create an energy self-sufficient development and provide the mechanism to become an energy provider to the surrounding area. Maximise use of efficient technologies such as wind and explore environmentally sympathetic tidal energy generation.

Aim to provide at least 50% of water requirements for flushing WCs and irrigation from rainwater harvesting to reduce the water consumption to 80 litres per person per day.

Flood Risk Management Approach

The LifE proposals aim to identify an adaptable plan for management of flood risk to the lower Arun that can respond to climatic changes, over and above current predictions. This requires a strategic short to long-term approach that considers the lower Arun within the context of the wider south coast.

Restoration of the Arun floodplain, increasing the river channel capacity and diversion of flood flows may form part of a strategic flood risk management plan.

For the harbour site additional planning measures should be applied to cope with the future flood risk.

Concept Ideas

The concept plan, shown in figure C22, shows two areas of development to the east and west of the site. Between these lies a large area of land set aside for flood storage, habitat, recreation. The location for the future river course is marked by low-lying wetlands.

A flood-resilient educational and activity centre is shown in yellow, uniquely located in the centre of the new habitat, on higher ground. The main concepts of the plan were:

- Integrate development proposals with long-term flood risk management strategy. Consider future diversion of the river and new constraints to reduce the pressure on existing defences of Littlehampton (after Royal Haskoning 2006).
- Recreate former tidal floodplain to provide flood storage, habitat and recreation.
- Divert flood flows around new development sites into dedicated flood storage lagoons.
- Use permanent water bodies to enhance economic opportunities through recreation and selective development.
- Prioritise development on higher ground, near existing settlements. Use contributions to pay for flood risk management measures.
- Use floodplain to locate large wind turbines.
- Divert tidal river water into some flood storage areas to provide daily tidal energy generation.
- Elevate and defend (as required) the Harbour/Westbank area to allow redevelopment of the site for the short to long-term, ensuring design is resilient to residual risk of overtopping of defences; and with the view to increasing future flood risk management measures to Littlehampton.
- Add to existing network of footpaths and cycleways to enhance river corridor and link to the coastline.

The Climping and Ford developments, located on higher ground, would be zero carbon but relatively traditional in their layout.

Harbour Site

Three options, shown in figure C23, were considered for the Harbour site. The 'land to water terraces' was thought to provide the most resilient and adaptable development solution for the harbour site, and offer river or lagoon views for every property.

Defences are provided but these are integrated into the layout plan, providing space for markets, skate parks and river walks.

Buildings are arranged perpendicular to the river course to avoid flood flows in the case of a breach of the river defences. Safe access would be provided by a partially raised arterial road, doubling as a flood barrier and by high level connections between buildings.

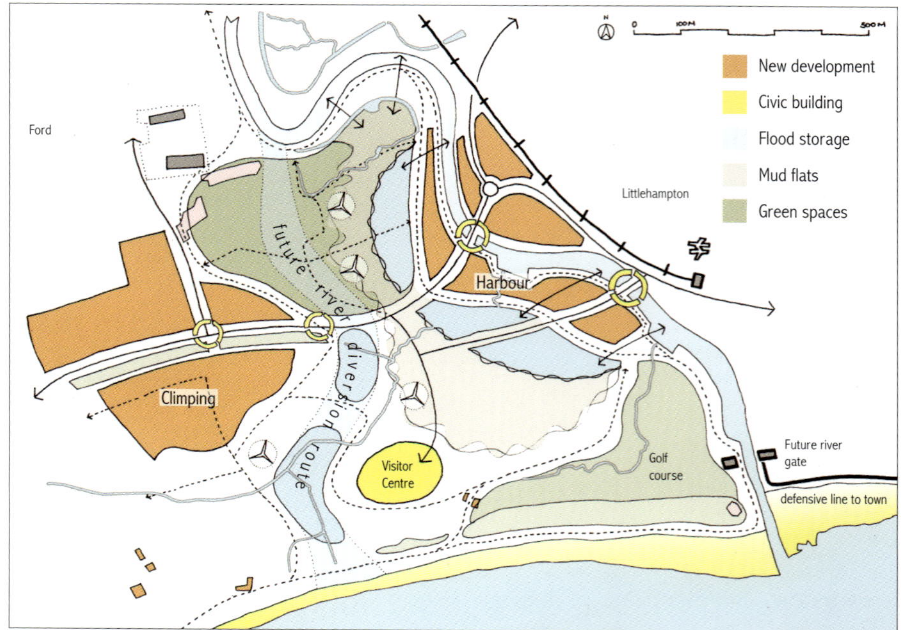

Figure C22 - Concept plan

'Spine Development'
raised central road with land stepped towards water

'Garden Bowl'
buildings raised above a garden bowl with strong defences

'Land to Water Terraces'
land steps towards the lagoon buildings

Figure C23 - Plan options for harbour site

A conceptual masterplan for the Harbour site was developed in combination with the large habitat-rich landscape, used for flood storage, energy generation and recreation. This is illustrated in figure C24.

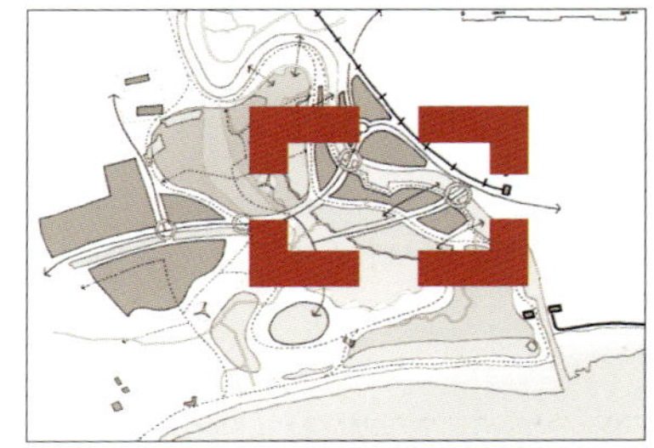

Tidal Lagoons – permanent water bodies created as part of the flood storage areas would be sheltered from the main river flow, providing opportunities for recreation and floating development.

Dynamic Habitats – a succession of habitats would be generated by the diversion of tidal water into the flood storage area on a daily basis. This would include: deep water, shallow water, inter-tidal mud and sandflats, saltmarsh, strandline debris, embankments, transitional habitats, grazing marsh and ditches.

Wet Windlands – large areas allocated for wetlands would allow 2 MW wind turbines to be sited away from development and infrastructure to avoid noise contamination and shadow flicker

Floating Village – floating houses, houseboats, barges, boats and pontoons could be moored alongside the developed edge of the tidal lagoons.

Floating Gardens – the lagoons could also provide space for floating gardens, used to produce local food as community gardens or allotments, or supporting facilities for storage of canoes and kayaks or changing facilities for swimming.

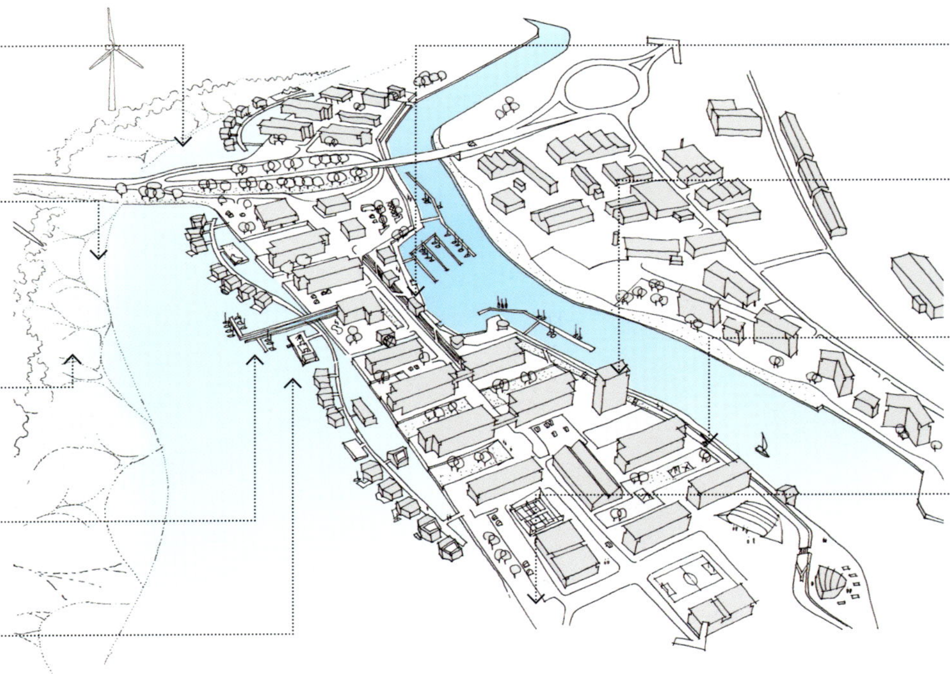

Figure C24 - Conceptual masterplan for Littlehampton

Promenade – riverside and lagoon edge sides would provide attractive waterfronts with small habitat strips, such as reed beds, public amenity play areas, gardens, walks and shelters.

Sky Gardens – private and public gardens could be provided on or above buildings, offering long views over the sea and landscape and providing rainwater run off attenuation to reduce risk of surface water flooding, particularly during high tides.

Green Defence Amenity – earth bunds, reinforced with stone gabions would be used to create wide terraced boundaries to the flood storage areas, allowing marginal habitat to establish and foot, cycle and bridle paths to be provided

Vertical Reedbeds – peripheral development edges could incorporate vertical reed beds, used to naturally treat and filter wastewater run off from anaerobic digesters then return to the watercourse.

Energy and Resources

The range of energy options available on the site allowed a far greater choice of technologies to be considered, including efficient commercial wind turbines and tidal turbines.

Three options were explored to identify the most cost effective solution to meeting demand and to demonstrate the viability of delivering zero carbon development without the need to use biomass.

Option 1 - Primary dependence on biomass CHP with multiple other energy sources

Option 2 - Primary dependence on biomass CHP with increased wind power to avoid need for solar PVs.

Option 3 - Primary dependence on wind power to generate electricity, which is then used to power heat pumps and electric heaters. Zero use of biomass.

Energy option 2 was found to be the most cost effective. It is also perhaps the most suitable solution for this site based on a balance of current tried and tested technologies combined with some innovation, through the use of tidal power. This option would allow further energy to be generated through use of solar PVs and additional wind turbines should they be required.

Energy option 3 is shown in Table C4 to illustrate the mix of energy sources proposed for a non-biomass dependant solution.

Technology	Yearly energy demand met *	Which demands met	Further info	% of total
Solar water heating	3,005 MWh	HOT WATER: 50% of all houses and flats	6,680 m² total active panel area required**	11.4%
Inter-seasonal thermal storage	1,195 MWh 1,195 MWh	SPACE COOLING: 90% of total SPACE HEATING: for thermal balance of system (35% total demand)	Aquifer thermal energy storage (ATES) assumed. Sizing for peak demand is inefficient, therefore 90% of yearly cooling demand only. For max efficiency, system should be in thermal balance.	9.1%
Ground source heat pumps (heating)	2,190 MWh	SPACE HEATING: Remainder of demand (65% total demand)	Approx 810, 70 m deep boreholes required to meet peak	8.3%
Tidal	1,345 MWh	ELECTRICITY: Max yearly production.	Contribution from FSA1 only taken account of at this stage. Assuming average 3 m usable head, 5 hour discharge time, 75% efficiency.	5.1%
Large-scale wind turbines	13,930 MWh	ELECTRICITY: To run ATES; to run GSHP; to run chiller to satisfy remainder of cooling demand; to satisfy remainder of HOT WATER requirement; to make up balance of demand	Approx 4, 2 MW turbines required.	66.1%

* Note that these figures have been rounded to the nearest 5 mWh

** Assuming flat plate collectors

Table C4 - Option 3 energy solution

Figure C25 illustrates the proportional contribution of each energy technology with each option. Figure C26 indicates the spatial demands of the energy technologies required for option 2, including the amount of land needed to grow biomass.

Biomass

The land area required to provide the volume of biomass fuel production identified in options 1 and 2 is 187 hectares (assuming the use of coppiced willow), 53% of the total site area. However, this land would be better used for flood protection and wood fuel could be provided from other local sources. The county council also stated that 'West Sussex is one of the most wooded Counties in the Country with an extensive resource of woodlands in need of management'. Therefore supply of biomass need not necessarily come from the immediate site.

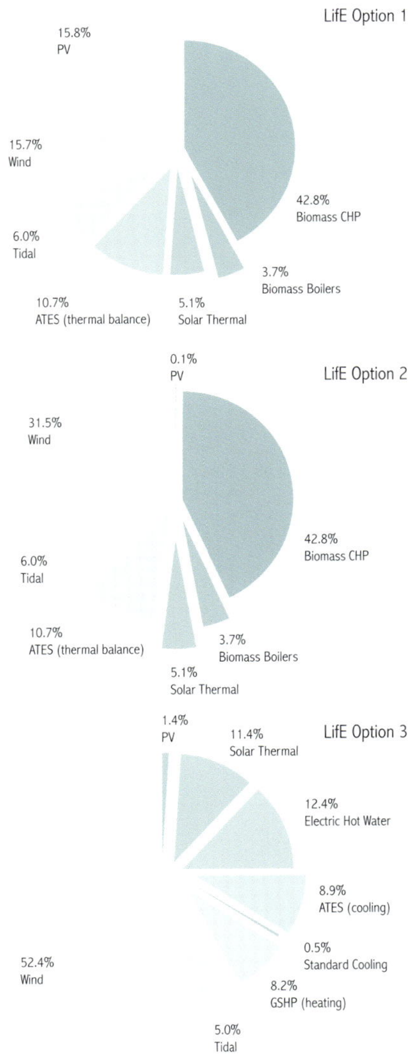

Figure C25 - Energy technology contribution to total demand

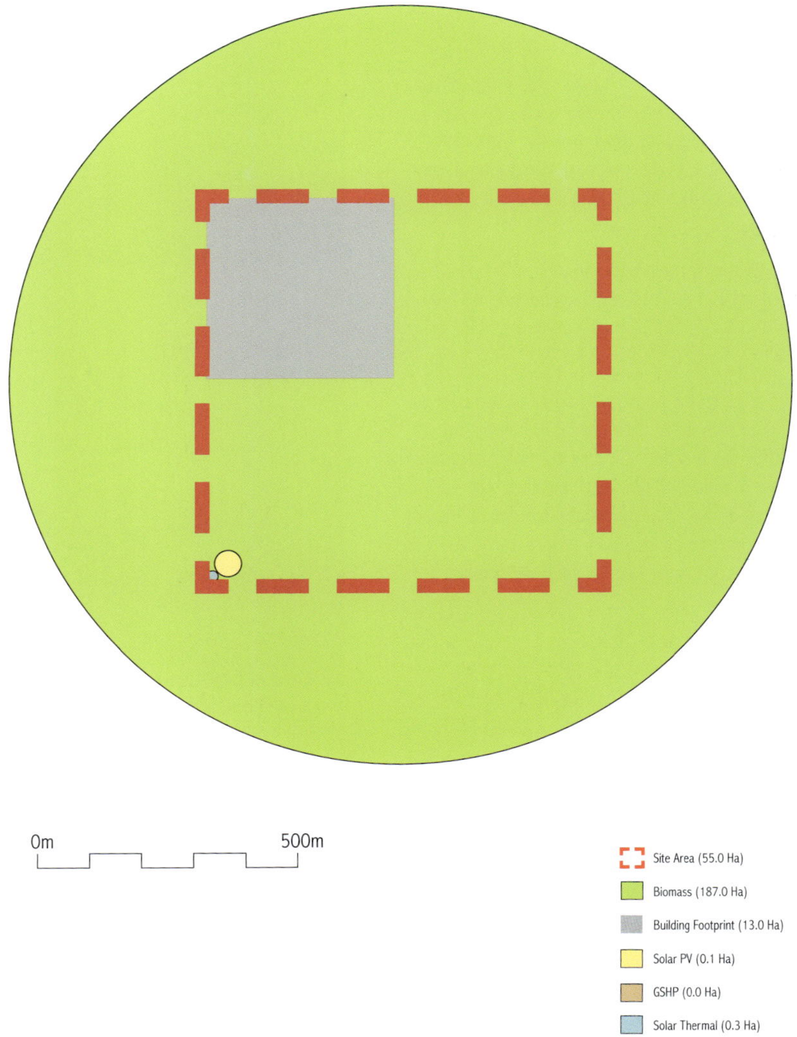

Figure C26- Space requirements for option 2

Energy and Resources

Wind

The potential to provide large-scale wind energy on the site was appealing due to the large area of undeveloped open land. Other considerations, such as roads, cycle routes, public rights of way and power cables may have a bearing on the siting of these.

An initial study based primarily on siting constraints suggests that the site could accommodate between four and six 2 MW turbines, see figures C27 and C28.

Theoretically this could provide all of the energy required to heat and power 2200 homes. In practice the electricity supply would be intermittent. However, wind turbines are typically linked into the national grid so that peaks and troughs in energy production are balanced by other energy sources to ensure no power loss to the development.

Figure C27 - 85 m high 2 MW wind turbines at Green Park Reading (top) and Ford, Dagenham (bottom)

Figure C28 - Location of wind turbines within the lower Arun site area.

Tidal Energy

A tidal stream or impeller (turbine) system located within the main river channel would affect the flow of the river, potentially affecting ecology and navigation.

An alternative system would be to transfer water from the river to a storage area and back again, driving turbines during the transfer, at the greatest tidal level difference.

The lagoons would fill up with water at high tide. This water would be retained then released at low tide, driving turbines on its way back into the river. Figure C29 shows two energy stations within the harbour area, each of which could house several turbines.

This system would disrupt the river less but would need a large water storage area such as a lake. It was proposed to use the flood storage areas and habitat for this purpose, with an area of permanent water forming an inland tidal lagoon.

Based on the approximate tidal curve (shown in figure C30) siting the discharge channel and turbine at -0.45 m AOD would allow approximately 4.5 hours energy generation during Neap Tide and approximately 5.5 hours during Spring Tide. These estimates are based on mean tides and do not take into account extreme river levels.

The power that could be generated from an array of tidal turbines using a 500 m x 1000 m (50 hectare) shallow flood store was calculated to be in the order of 1,300 MWh per annum. Use of further flood storage areas might generate in the order of 3,000 MWh, similar to the amount of power produced by one large 2 MW wind turbines.

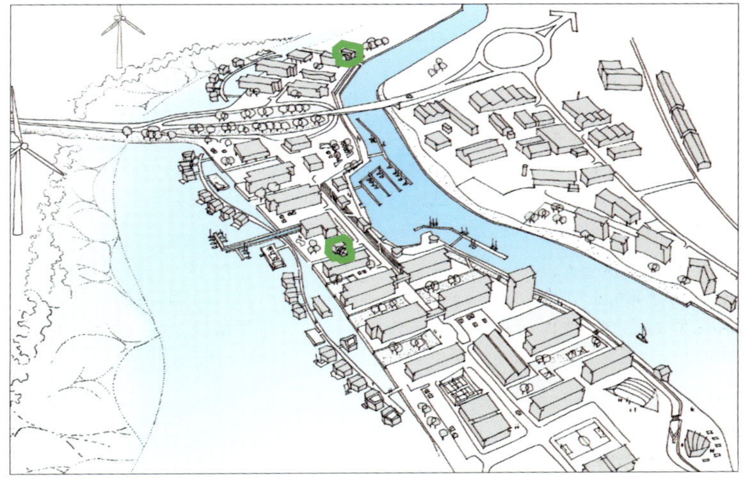

Figure C29 - Location of tidal power stations within the harbour development

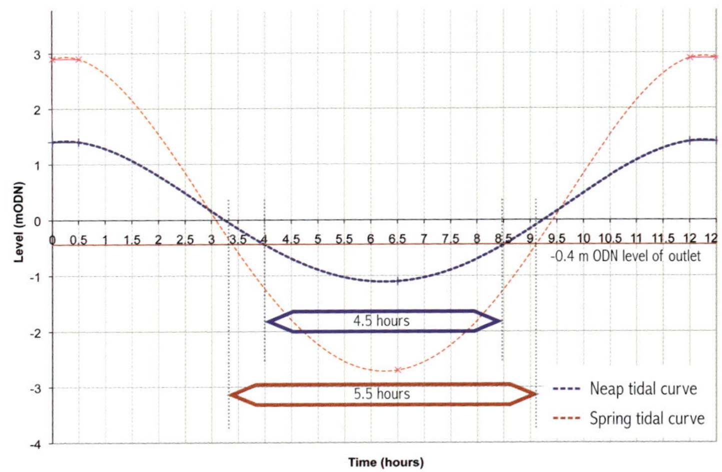

Figure C30 - Littlehampton UNA Wharf - approximate tidal curve and potential power generating periods

Energy and Resources

The Water Solution

To reduce the water consumption from 110 litres per person per day (the level of consumption set by the Code Level 3) to the target 80 litres per person per day set by the Code Level 6. A further saving of 30 litres per person per day would need to be provided from rainwater harvesting.

If the number of consecutive dry days is likely to be a maximum of say 20 the total required harvesting would be approximately 3,300 m^3. If it is only five days then 825 m^3 storage must be provided.

On the basis of five days, seventeen 50 m^3 tanks would be required within the Harbour site to achieve sufficient capacity to meet Code Level 6.

Figure C31 shows a communal rainwater harvesting tank located between each building to serve the needs of the surrounding occupants. These would be between 45 m^3 and 55 m^3 and connected to the gravel swale SUDS system. These tanks would supply over half the local water flushing and irrigation requirements, with surplus storage allowing capacity during flooding.*

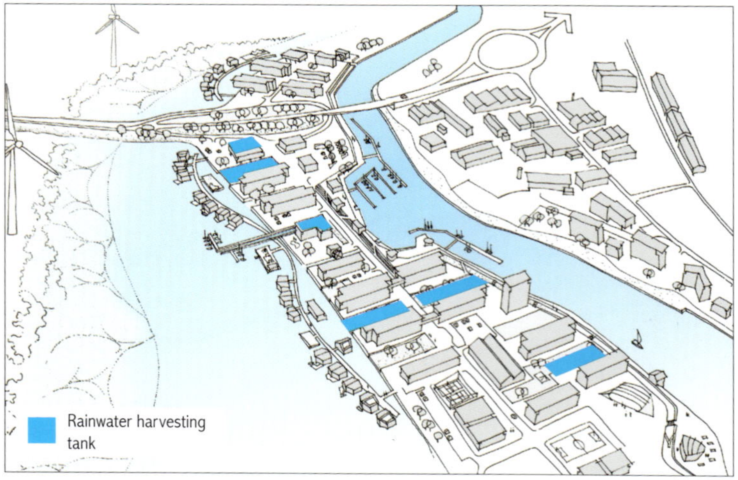

Figure C31 - Rainwater harvesting areas

* No storage amount was added to the total SUDS calculation on the basis that the rainwater harvesting tank could be full at the time of a storm.

The 1st Review of the Shoreline Management Plan[48] (awaiting Defra approval at time of writing) identifies 'managed realignment' as the policy for the coastline between Littlehampton Harbour/ Westbank and Poole Place. At Littlehampton the policy is to maintain the existing defences 'hold the line'. This is the approach for the medium term (50 years) and long-term (100 years).

At Climping the policy is to reduce the future requirement for beach recycling (from east to west) and retreat the shoreline to a more stable position. Additionally the River Arun training wall should be replaced when it fails in about 25 years.[47]

The policies for the lower sections of the Arun are a combination of 'hold the line', 'managed realignment' and 'defence raising' (Halcrow, 2007). However, raising the defences on the east bank of the Arun is constrained by the existing development in Littlehampton and likely to prove very expensive.

An over-riding objective in setting present day Flood and Coastal Erosion Risk Management (FCERM) measures is to work with natural processes and avoid the construction of 'unnecessary' flood defences, particularly where these defences would remove areas of fluvial or coastal floodplains.

The flood risk management objectives are listed below:

- Integrating Flood Risk Management into strategic planning.
- 'Sea level rise linked regional planning' with relocation of vulnerable communities.
- Large-scale flood storage to reduce flood risk to new and existing developments.
- Reinstate the natural tidal floodplain where possible, reducing costs of defending low value land.
- Managed realignment of coast to higher ground to reduce wave energy on realigned defences and to provide a more sustainable coastal bay shape.
- Locally improve and raise defences to existing vulnerable properties and areas and key infrastructure.
- Develop resilient properties able to withstand residual risk posed by extreme storm events.
- Reduce risks to communities by ensuring they understand the flood risk management mechanisms and evacuation procedures.

Restoration of the natural floodplain should enable a reduction in flood risk and could provide habitat (ie. gravel beaches, vegetated shingle, dunes, transitional inter-tidal habitat, regulated tidal exchange, salt marshes, mudflats, meadows, etc.).

Regional Flood Risk Management and Development

The Arun Riviera

The first aim was to establish a regional strategy for development for the long-term management of flood risk in the area. This was based on examining the effect of climate change, in particular sea level rise, on the site and the surrounding area.

An incremental regional strategy based on sea level rise scenarios was developed that explored the restoration and management of the former natural floodplain of the Arun, such that it could be used as flood storage during extreme events.

With higher sea level rise scenarios the lowest section of the river could need re-routing and widening to maintain river flow in and out (after Royal Haskoning[49]).

Indicative development sites were proposed along the edge of the river, further up towards Arundel and the concept of creating an 'Arun Riviera' was developed (see Figure C32a to c).

The Arun Riviera concept could be used in the current development plan, identifying potential sites along the banks of the floodplain. The Riviera might extend down to the current coastal limit but on the basis that these settlements in the low-lying land could need to be relocated in the future. This could mean that the rights to occupy a site are sold rather than the freehold for the site - 'sea level rise linked land leases'.

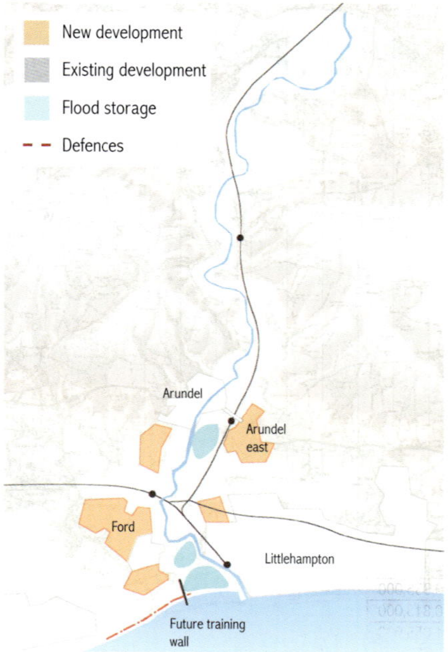

Figure C32a - Exploratory development plan for up to 1 m Sea Level Rise

Figure C32b - Exploratory development plan for up to 3 m Sea Level Rise

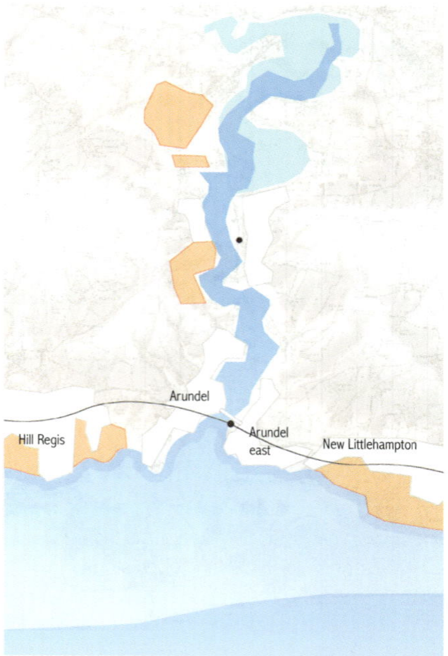

Figure C32c - Exploratory development plan for over 6 m Sea Level Rise

Note:

These plans are an indicative and high level interpretation of the possible effects of long-term climate change on the area. They do not reflect any scientific study. Current predictions are that sea levels will rise by up to 1 m by the end of the century.

The Lower Arun

The first stages of the regional strategy would be to reconnect the river to the tidal floodplain. This could be implemented at the lower end of the Arun, near Littlehampton, to trial effectiveness prior to recreating floodplain along further up the river.

The volume of water within the river channel was calculated based on the mean flow rate at the mouth of the river and the width of the channel. Several channel widths were explored to identify the implications of re-routing and widening the river mouth.

Flood storage areas and their corresponding capacities were identified along an 18 km stretch of the lower Arun*. These were added to the capacity of the river channel itself to determine the number of storage areas required for a given storm event.

Figure C33 shows the potential storage areas identified within the river valley. Table C5 shows the flood storage areas required for given return periods including three sea level rise (SLR) scenarios and future river mouth widening from 40 m to 60 m (to reduce pressure from rising water levels on existing defences).

The (untested) results suggest that the volume of water for a 200 year flood with 0.5 m sea level rise could be contained within the river channel and three storage areas within the site area.

Widening the river mouth showed that greater water volumes would enter the river, reducing the effectiveness of the flood storage areas, therefore requiring additional storage to be provided.

Flood Storage Capacity Required			
Event		Volume (m^3)	Reference
MHWS	Tidal flood volume - contained in exg system	3,859,200	A
1 in 20	Tidal flood volume - contained in exg system	4,924,800	B
1 in 200 +0.5 m SLR		6,076,800	C
1 in 200 +1.0 m SLR		7,336,800	D
1 in 200 +2.0 m SLR		9,136,800	E
1 in 200 +2.0 m SLR	Assumed entrance widened to 60m	13,705,200	F

Flood Storage Area Ref. No.	Length (m)	Width (m)	Depth (m)	Volume (m^3)	Cumulative Capacity (m^3)	Return Period Storage Demand Met
River Channel	18000	50	5.5	4,950,000	4,950,000	(B) 1 in 20yr
1	1000	500	1	500,000	5,450,000	
1a	600	200	1	120,000	5,570,000	
2	900	600	1	540,000	6,110,000	(C) 1 in 200yr incl 0.5m SLR
3	500	400	1	200,000	6,310,000	
4a	300	300	1	90,000	6,400,000	
4b	300	700	1	210,000	6,610,000	
4c	600	200	1	120,000	6,730,000	
5a	500	400	1	200,000	6,930,000	
5b	2000	400	1	800,000	7,730,000	(D) 1 in 200yr incl 1 m SLR
6a	1200	700	1	840,000	8,570,000	
6b	500	500	1	250,000	8,820,000	
6c	500	500	0.5	125,000	8,945,000	
6d	250	400	1	100,000	9,045,000	
7	-	-	1	690,000	9,735,000	(E) 1 in 200yr incl 2 m SLR
8	500	600	1	300,000	10,035,000	
9	500	600	1	300,000	10,335,000	
10	1200	400	1	480,000	10,815,000	
11	800	300	1	240,000	11,055,000	
12	700	350	1	245,000	11,300,000	
13	500	300	1	150,000	11,450,000	
14	250	250	1	62,500	11,512,500	
15	500	150	1	75,000	11,587,500	
16	1600	400	1	640,000	12,227,500	
17	800	300	1	240,000	12,467,500	
18	800	700	1	560,000	13,027,500	(F) 1 in 200yr incl 2 m SLR

Table C5 - Outline flood storage calculations based on extreme water levels and existing ground levels.

* This was a high level exercise to explore the potential for flood storage within the Arun and no detailed modelling was carried out.

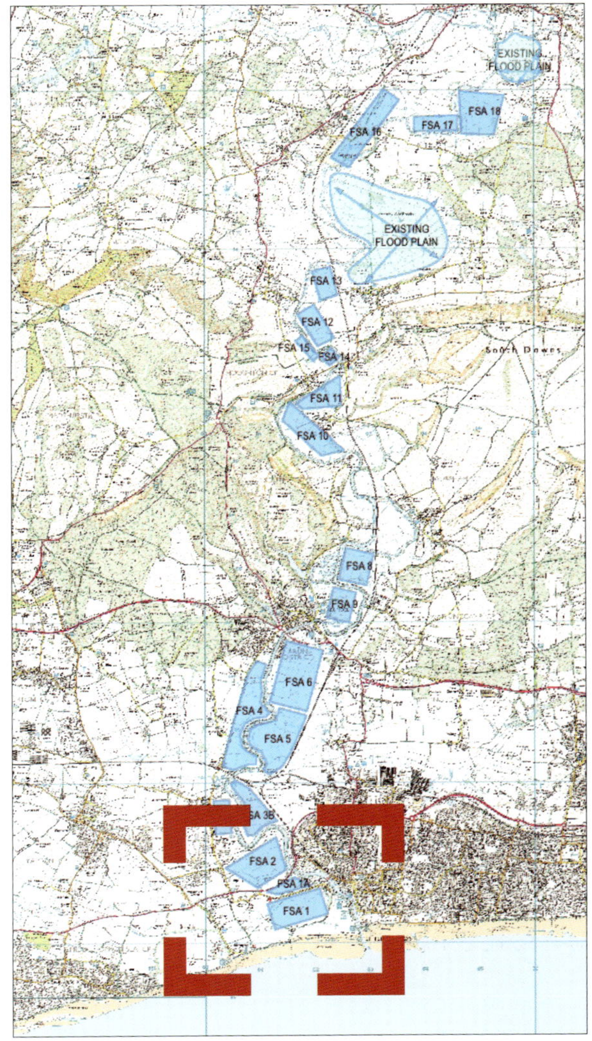

Figure C33 - Flood storage areas

Flood Risk and Development

Further to the flood risk management proposals for the region, a strategy was developed for the development sites to manage the remaining flood risk.

Following the guidance of PPS25 the site proposals were considered on the basis of a risk-based approach and the sequential test. Therefore the initial approach was to allocate development to areas of lowest probability of flooding.

Site Specific Development

Figure C34 indicates the flood zones for the site. The main proposed development area is to the west of the site at Climping/Ford on the highest ground in Flood Zone 1. Based on the PPS25 guidance on vulnerability of development use, all forms of development would be appropriate.

The majority of the area is not identified for development, with the exception of a small area to the south, which is within Flood Zone 2 and would be used for a summer wetland visitor centre and refuge.

Some of the current site uses would be relocated to avoid risk. This would include relocating the caravan park and the few buildings located in the most vulnerable areas.

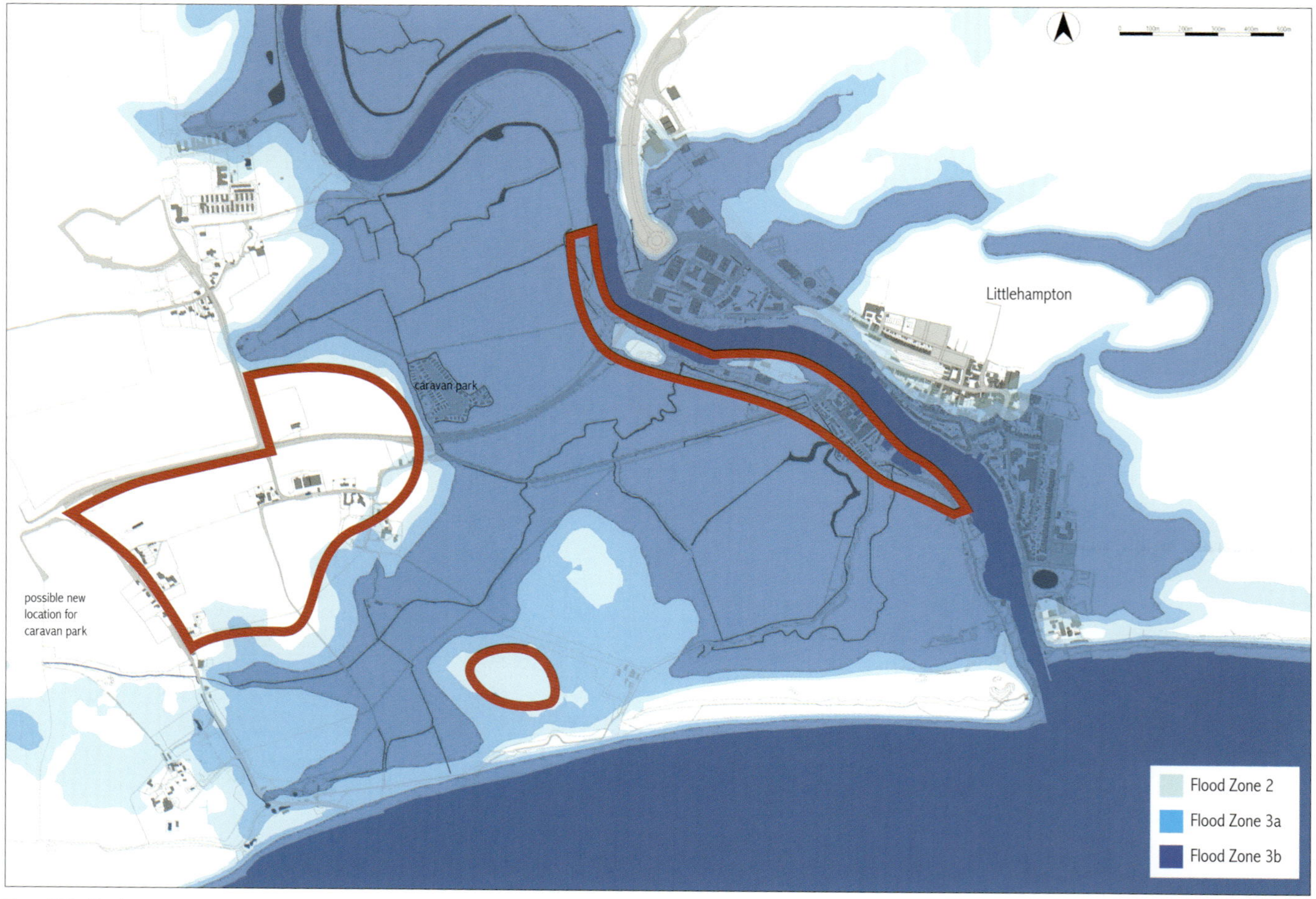

Figure C34 - Flood map, current

The Harbour/Westbank Site

Large parts of the Harbour/Westbank site to the east were found to be below 3.8 m AOD (above ordnance datum) and therefore lie within Flood Zone 3b, classified as 'Functional Floodplain'. The only appropriate uses would be water compatible development, such as shipbuilding and repairs, docks and marinas*. However, this site is an important site for the regeneration of Littlehampton and in this context there are no other sites available. To consider the site for other forms of development the land must be raised above 3.8 m AOD so that it is no longer within Flood Zone 3b (figure C35). Compensation for the loss of floodplain may be required and the development subject to the exception test**. The exception test is divided into three parts, these are:

Demonstrate wider sustainability benefits

Development of the site was seen as integral to the flood risk management of the region. The development would help provide economic investment in Littlehampton, provide new homes and businesses, improve management of flood risk for existing residents, provide the future flood defence line to the town (if the river channel was re-routed), provide energy generation, and remove possible contaminants on the site.

Development should be on developable previously-developed land (brownfield)

The Harbour/Westbank site has been used for boatyards, houses and light industry.

Demonstrate that the development will be safe, without increasing risk elsewhere, and, where possible, will reduce flood risk overall***

Figure C36 indicates the potential sources of flood risk to the site.

The following pages describe some of the flood risk management measures explored. A Flood Risk Assessment is provided, later in Table C7.

* In accordance with PPS25 Table D.2 – Flood Risk Vulnerability Classification.

** The Environment Agency has determined that compensatory storage is not required for sites where the flood risk is tidal.

*** Demonstration that the development will be safe is assumed to be required for the lifetime of the development, which is assumed to be 100 years.

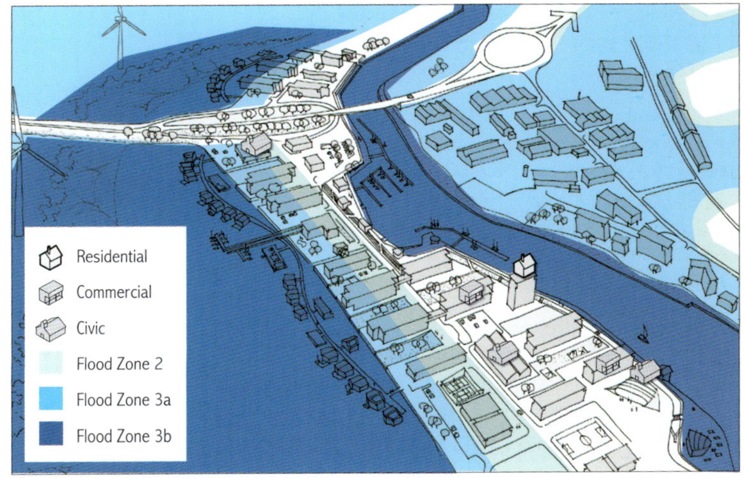

Figure C35 - Flood zones and development*
Note flood zones after development are indicated

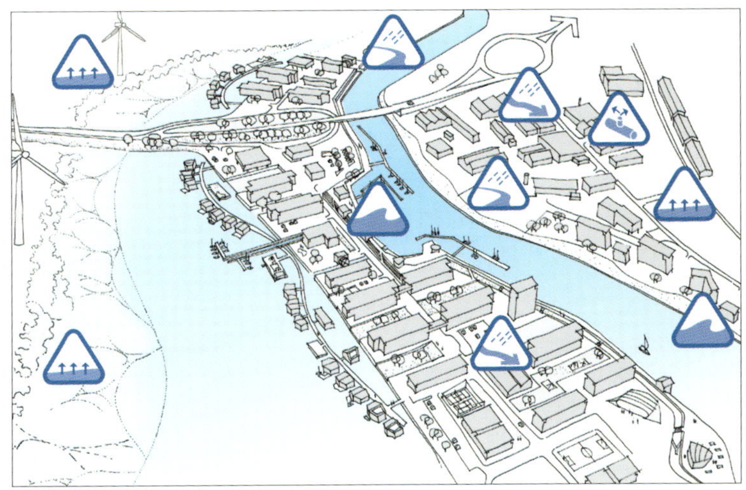

Figure C36 - Potential sources of flooding

Flood Risk Management

To demonstrate that the development will be safe the severity of the flood hazard must be identified such that an appropriate solution/s is identified for the lifetime of the development.

A number of solutions would be required to manage: the primary risk of flooding from rivers and the sea; the secondary risk of flooding from groundwater, land and sewers; and the residual risks after the proposed flood risk management solutions are implemented.

The development of the large flood storage areas (as part of regional improvements) are likely to reduce flood levels to Littlehampton and Arundel, provided that the existing tidal constraint at the mouth of the river is maintained (to restrict the volume of water entering the river).

Land would be raised by 0.3 m to 1.0 m to elevate the ground level above the 5% (1 in 20 year) probability flood level and equivalent to Flood Zone 3a (possibly reducing pressure from groundwater flooding). Some areas of land would also be raised above the 0.5% (1 in 200 year) and 0.1% (1 in 1000 year) flood levels to Flood Zones 2 and 1 respectively.

Raised embankments set back from the existing river edge would be designed to cope with extreme flood levels, wave action, and provide scope to be increased if sea levels rise higher than predicted.

Site planning and buildings would be designed to cope with residual risk of failure of embankments and/or flood storage areas. Design measures included the following:

- Lower vulnerability uses, such as shops, cafes and businesses located on ground floor
- Flood resilience. Buildings would be constructed with concrete frames to reduce the possibility of structural damage that could be caused by deep (over 0.6 m) floodwater carrying debris.*
- Buildings located back from river bank defences and orientated perpendicular to defensive walls.
- Secondary access routes are provided at high level. These are designed as generous timber decks that provide shelter from rain and sun to the retail units below (see Figure C38).
- Level variations would create flow buffers to slow the water and temporarily contain it within specific areas, allowing more time for people to evacuate.

Safe haven buildings (shown in figure C37) built to a higher standard than surrounding properties could provide sanctuary from existing properties and help facilitate recovery after a flood.**

* If a floodwall fails deep water can rush through at speed. This could cause structural damage to buildings, as seen in New Orleans in 2005.

** Safe access throughout the duration of a flood is normally a requirement of the EA. However, it is unlikely that individuals will seek to leave their premises during a storm surge, unless forced to do so. In this situation it may be better to consider riding out the storm in a well protected home or safe haven until it has passed.

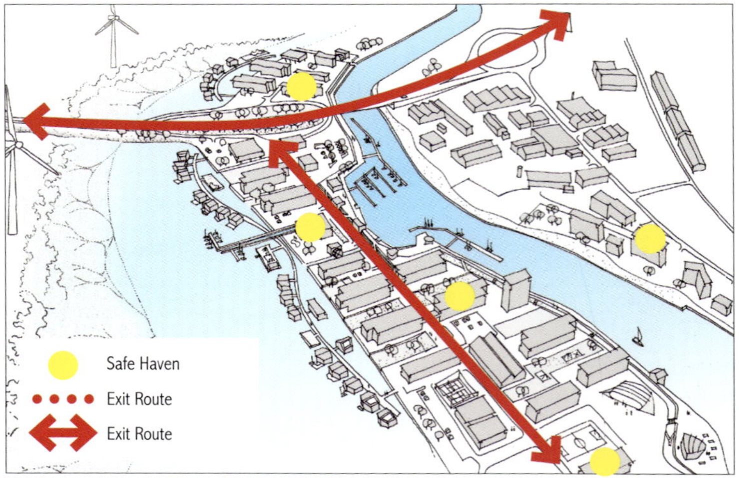

Figure C37 - Safe havens, evacuation routes and flood stores

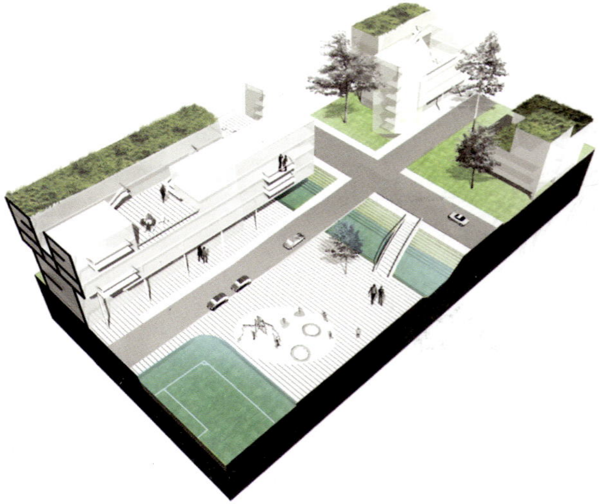
Figure C38 - Residential units with elevated access constructed above retail and workshops

Sustainable Drainage

A Sustainable Drainage System (SUDS) would be used to cope with the 'Pluvial Flood Risk', caused by high tides preventing conventional drains from discharging rainwater water.*

The aim of a SUDS system would be to attenuate all storm water on site for a duration equivalent to the time required for the tide to retreat. Because the site would be defended, SUDS measures could be located at ground level. However any breach of the defences would render the SUDS system ineffective and therefore measures located above ground level, such as green and brown roofs and green walls would be preferable.

Large underground communal rainwater harvesting tanks would collect and reuse rainwater, providing some small additional storage.

Localised impermeable drainage parks would store rainwater away from homes and prevent ingress of groundwater flooding (see Figure C39).

Permeable paving and swales were not considered appropriate due to the likelihood of groundwater flooding and the potential for contaminants on site.

An anaerobic digester, located above ground level, reduces material entering the sewers therefore reducing risk of flooding further and can be combined with biogas generation.

* Failure of drainage pumps following the initial flooding of New Orleans in 2005 resulted in significant pluvial flooding after heavy rainfall was not pumped out of low-lying land.

Because the River Arun is prone to flash fluvial flows in the winter the river level can rise and fall quickly. This can result in river levels higher than drainage outlets, preventing rainwater from draining from the site. This can also occur during a high tide or storm surge. In this event rainwater would need to be pumped into the already high river. Therefore attenuation of rainwater on site is an important feature of flood risk management.

The Environment Agency policy is to replicate 'greenfield' surface water drainage rates where possible. The local geology, depth to groundwater and contamination would result in an infiltration system being largely ineffective and unsuitable. Therefore the SUDS solution is to install source control measures at high level to attenuate the run off, with rainwater storage integrated with impermeable sunken play areas, designed to resist groundwater surge.

A single Multi-use Gaming Area (MUGA), sized to accommodate three basketball courts and set 1 m below the surrounding ground level would provide approximately 2000 m^3 storage. Therefore rainwater storage is provided through a combination of sunken play areas, green roofs and some additional capacity from communal rainwater harvesting systems (see figure C37).

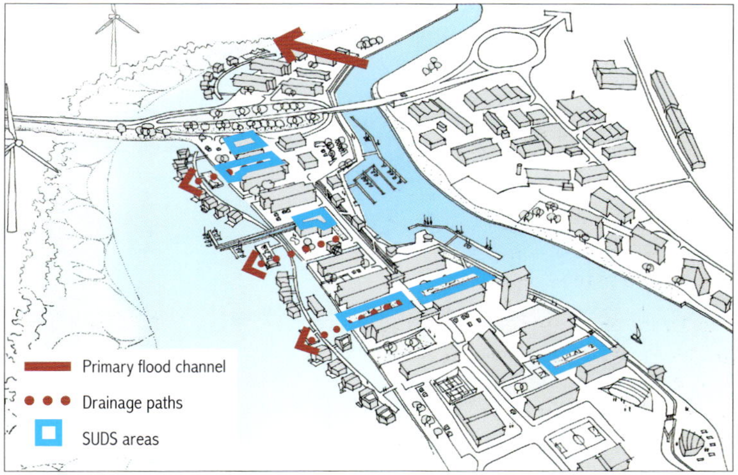

Figure C39 - SUDS areas and drainage paths

Landscape

The majority of the site would be set aside for intertidal habitat and grazing marsh. As the tide rises areas of land would be covered by water and exposed as the tide recedes. This provides a range of inter-tidal habitats at different land levels for flora and fauna. Habitats would include gravel beaches, vegetated shingle, dunes, transitional inter-tidal habitat, regulated tidal exchange, saltmarshes, mudflats and meadows.*

A simple assessment of the potential areas for inter-tidal habitat, based on typical tide levels and existing ground levels is shown in figure C40.** Figure C41 shows how this would change with a metre of sea level rise.

- The areas in pink and orange indicate potential areas of mudflat.
- The areas in light green indicate potential areas of salt marsh
- The areas in dark green indicate transition habitats
- The areas in white are above typical high tide levels
- The areas in blue are wet throughout most tidal cycles

This landscape would provide varied character and interest, which would provide different opportunities for access throughout the day.

Figure C42 indicates the types of plant species that may be supported by these habitats.

* Halcrow 2007

** A detailed assessment would need to be carried out to take account of wave action, high river flows, ground/soil conditions etc. This assessment is based on CIRIA, 2004, Coastal and estuarine managed realignment design issues[50] and assumes that the potential area of inter-tidal salt marsh is between MHWS and mean high water neap tide level (MHWN); and the potential area of inter-tidal mudflat are between MHWN and MLWS.

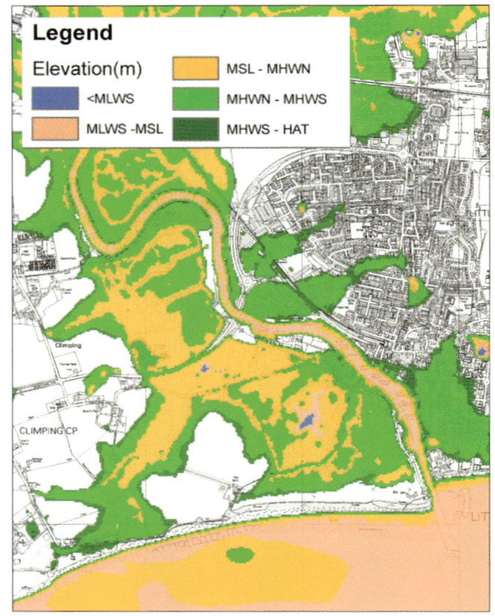

Figure C40 - Potential Habitat Creation Map

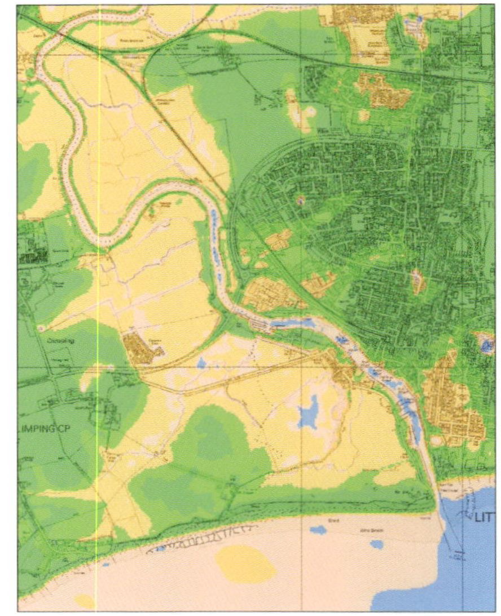

Figure C41 - Potential Habitat with 1.045 m sea level rise

Figure C42 - Potential saltmarsh plant species

Amenity

In addition to the large open recreation areas a number of local play amenities will be provided within the development sites, shown in figure C43.

Figure C44 shows a play area designed to accommodate and withstand flood water.

On the Harbour/Westbank site the following play areas would be required to contribute to the total amenity demand:

- 1 no. 1000 m^2 NEAP
- 2 no. 400 m^2 LEAPs
- 20 no. 100 m^2 LAPs
- 4 no. 500 m^2 MUGAs
- 20 no. 250 m^2 Allotments

Precedents

Figures C45a to C45d show examples of other sites where similar ideas have been implemented.

Plans and Sections

Figure C46 shows an indicative development plan for the Harbour site. The buildings are orientated perpendicular to the river to minimise resistance to flood flows if the river defences fail, also allowing views of the river from all apartments. The space between the buildings may be used for public gardens, recreation and bio-retention SUDS.

Figures C47a to C47d show a section through this plan.

Figures C48a to C48c illustrate an area of the Harbour site before and during rainstorms.

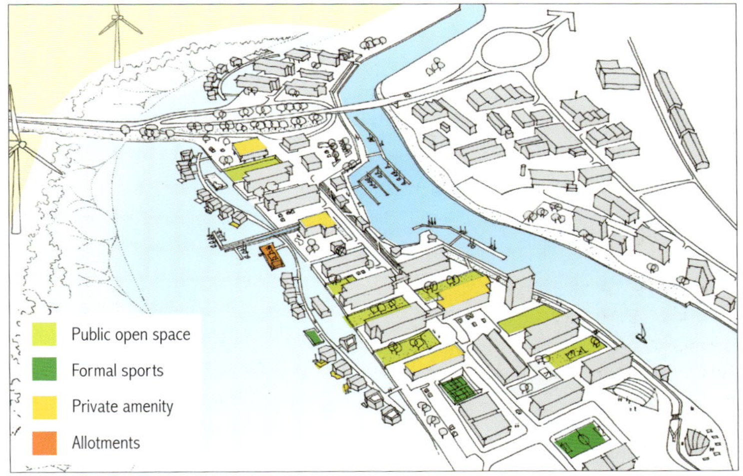

Figure C43 - Landscape areas

Figure C44 - Flood resilient play area

Precedents

Figure C45a - Moorings on the River Welland, Lincolnshire

Figure C45b - Amphibious and floating waterside development in Maasbommel, the Netherlands

Figure C45c - Dordrecht floodproof pilot project

Figure C45d - Tilbury marshes, Thames gateway, Essex

'Floating Village'
- Gardens and allotments
- Pontoon walkways
- Floating lidos
- Recreation buildings

'Causeway Roads'
- Planted banks
- Safe passageway

'Dynamic Habitats'
- Mudflats
- Saltmarsh
- Transitional habitats

'Grazing Marsh'
- Livestock grazing
- Occasional flood storage

'Large Wind'
- 500 m from nearest development

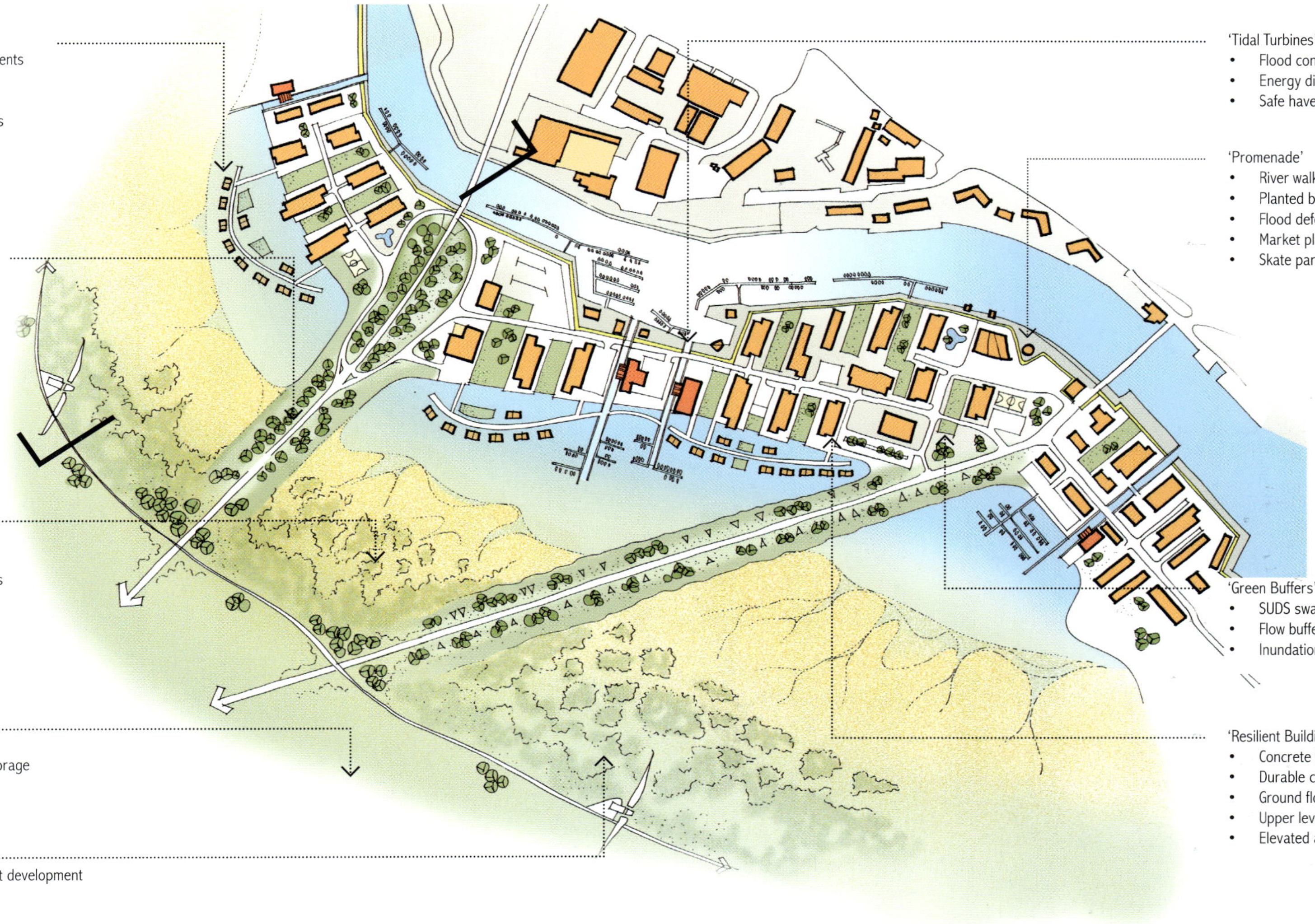

'Tidal Turbines'
- Flood control channels
- Energy distribution centre
- Safe havens

'Promenade'
- River walks
- Planted banks
- Flood defence
- Market places
- Skate parks, bars and activities

'Green Buffers'
- SUDS swales and channels
- Flow buffer roads
- Inundation control areas

'Resilient Buildings'
- Concrete frame
- Durable construction
- Ground floor retail and workshops
- Upper level residential
- Elevated amenity and access

Figure C46 - Site plan

Sections

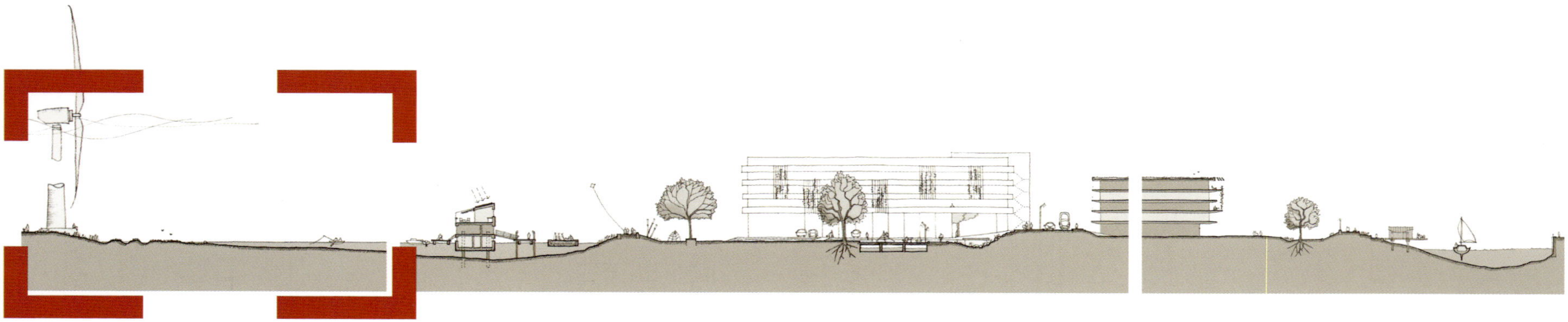

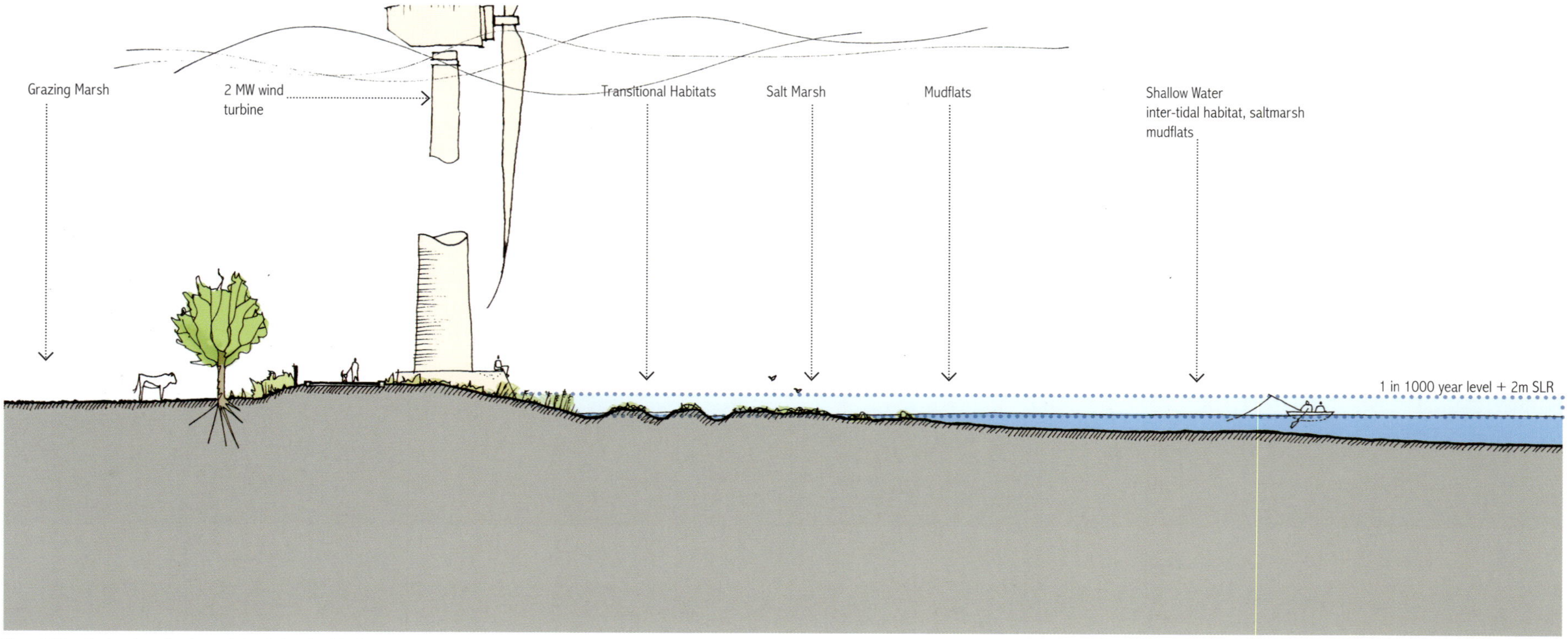

Figure C47a - Section through salt marsh (vertical scale of landscape exaggerated)

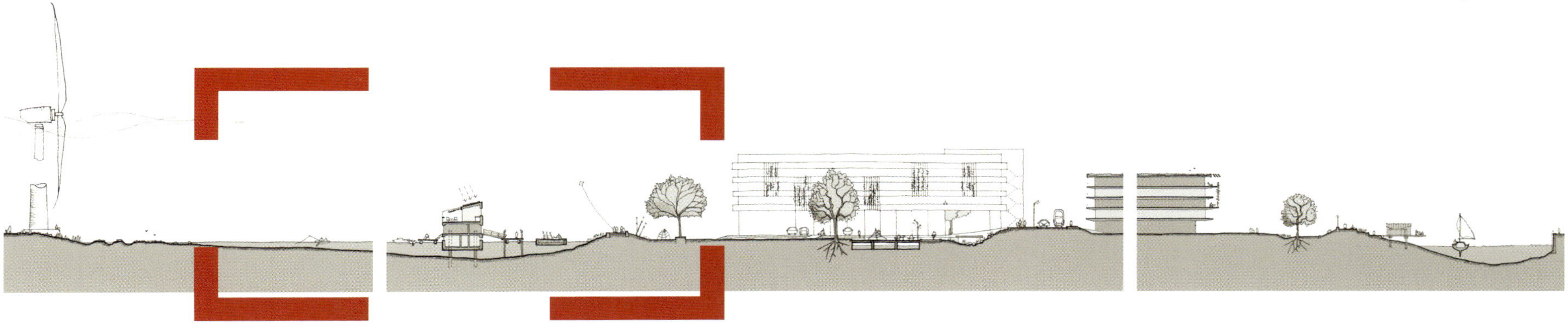

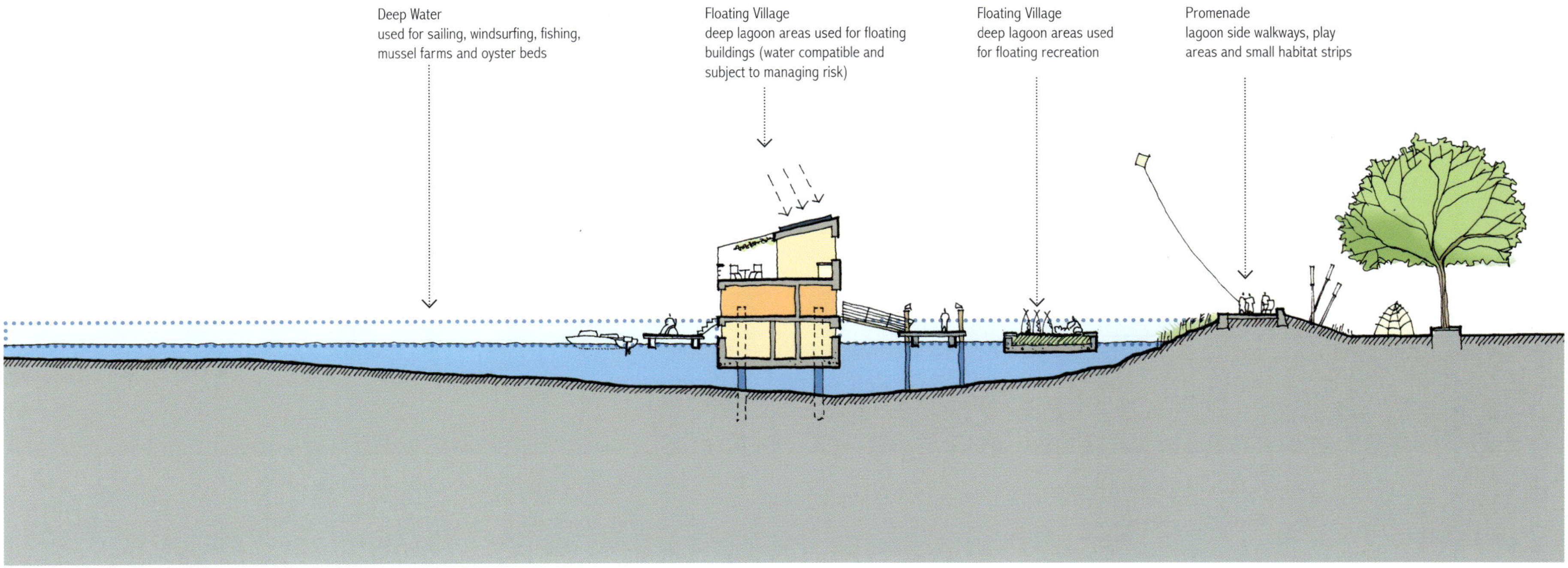

Figure C47b - Section through tidal lagoon (vertical scale of landscape exaggerated)

Sections

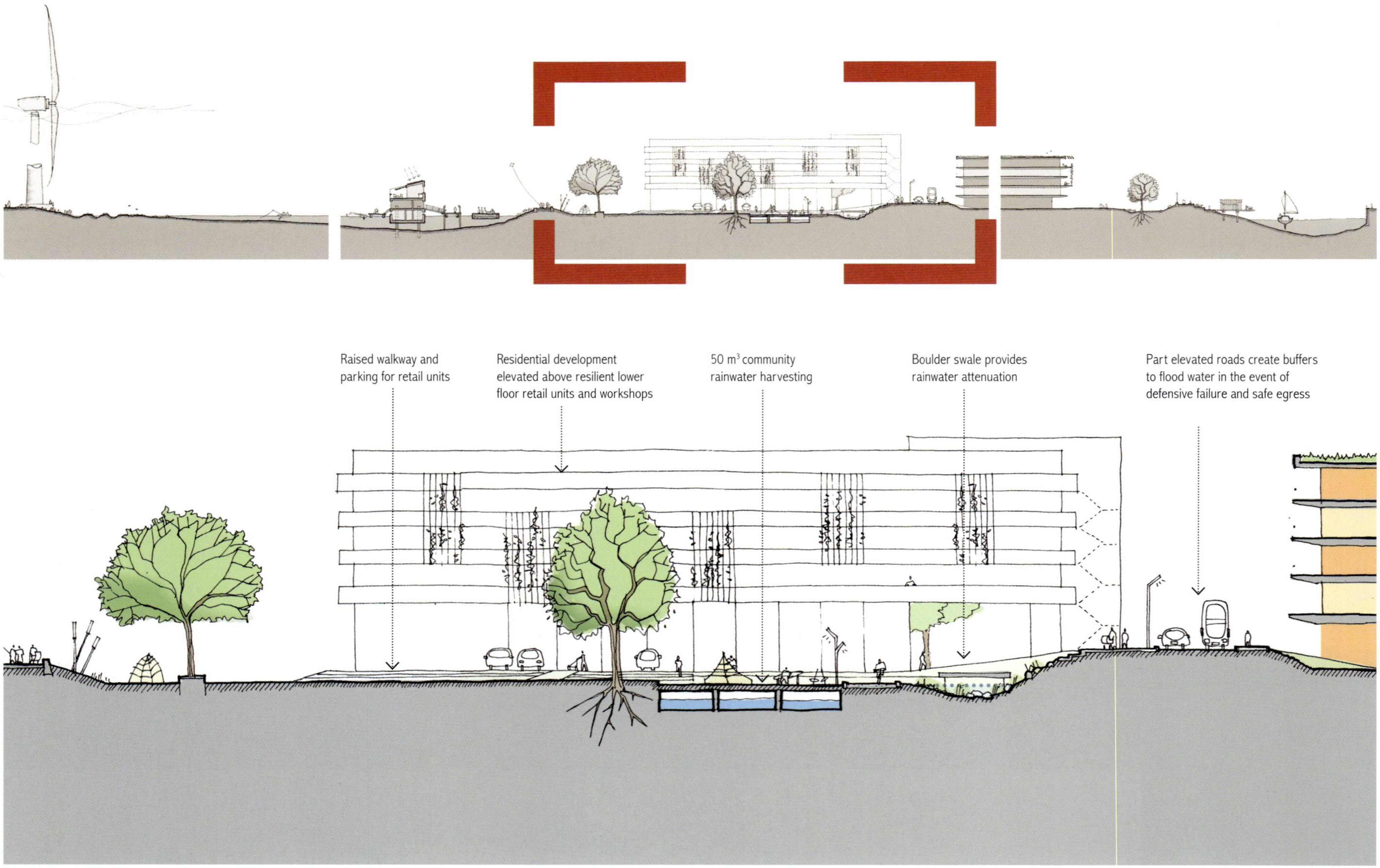

Figure C47c - Section (vertical scale of landscape exaggerated)

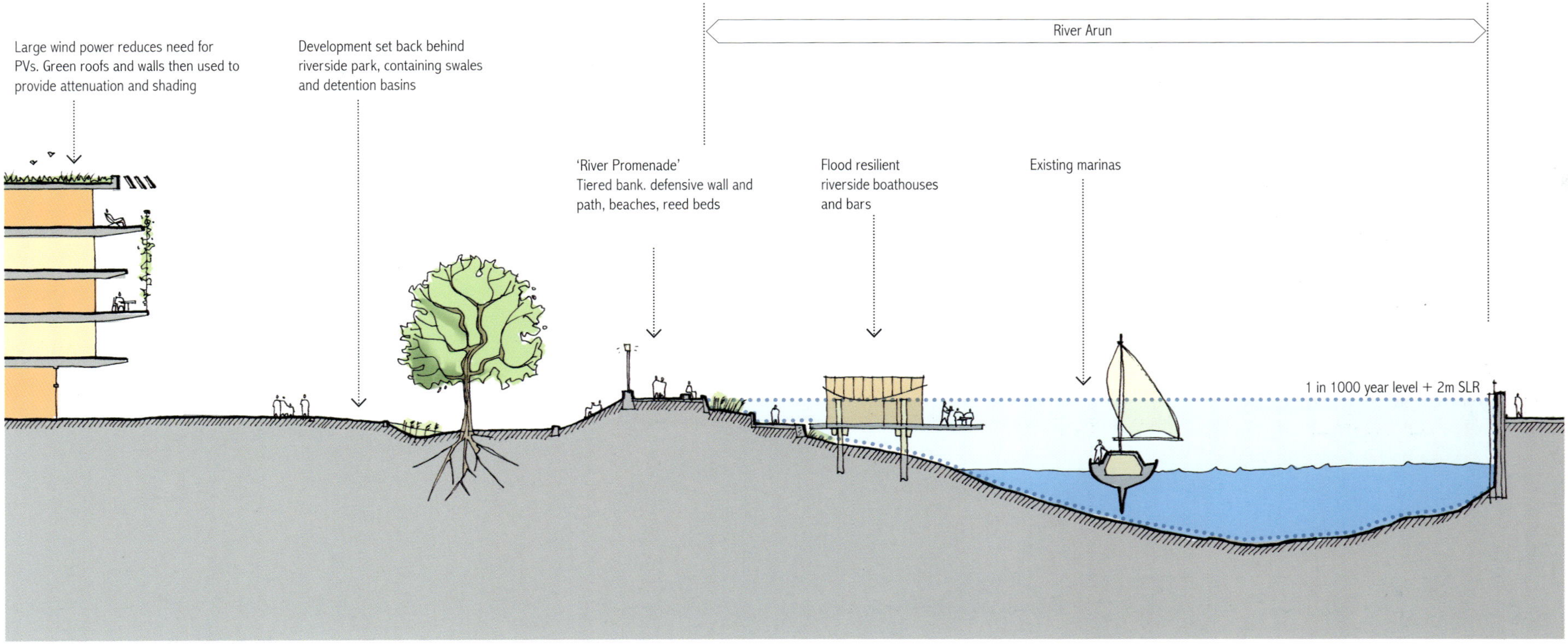

Figure C47d - Section through the River Arun

Figure C48b - During a 1 in 20 year storm rainwater is retained in rain courtyards, reducing run off overland and into the sewers.

Figure C48c - During a 1 in 200 year storm peak water is diverted into the distant lagoon, reducing river levels. The courtyards retain rainwater that cannot drain into the river and in the unlikely event of a river wall breach, high pathways allow safe access to property.

Figure C48a (left) - View, showing linear development, boulder swales, promenade and distant estuarine lagoon.

Review >

The design proposals for all three case studies are assessed collectively in the main report (pages 26-29). A summary of the main issues, particularly with regard to flood risk management, is given below.

Costs

Due to the extreme nature of the coastal conditions the cost of flood management is much greater than the other two study sites. On the basis of providing 2200 new homes built to current building regulations, flood risk management costs amount to 11% of the total cost as opposed to 1% or 2% on sites 1 and 2.

Table C6 shows a breakdown of the costs for the three energy options.

Because proposed flood risk management measures should benefit neighbouring communities at Littlehampton, Ford and Arundel the total costs are unlikely to be paid for solely by developers.

When an integrated approach was applied, as advocated in the LifE options, the cost of flood management reduced by almost 20%, to 9% of the total cost. However, this does not include any costs associated with land purchases or grants for habitat generation.

The most significant cost increase of the LifE options over the Base Option was the energy and utilities. However, this only amounted to 11% of the total development cost as opposed to 20% to 22% for sites 1 and 2. This was primarily on the basis that large land-based wind turbines could be used.

Overall the LifE option 2 offered the most cost effective solution, requiring an increase of only 11% over building to current standards.

Over £6m may be available for the creation of habitat on the site, either through government grants or from proposals that must provide compensation for loss of habitat.

The opportunity to create a second marina within the tidal lagoons and property with riverside views or lagoon views would also significantly improve the sales receipts for the site.

Site 3, Development Costs

	Base Option Costs	LifE Option 1	LifE Option 2	LifE Option 3	Percentage change from base (Option 2)
Construction	£259,943,090	£287,341,052	£287,291,612	£287,544,992	10.52%
Transport and Land Use	£90,050,885	£71,897,435	£71,961,035	£71,443,515	-20.09%
Flood Management	£53,028,400	£45,017,200	£44,850,880	£44,941,780	-15.42%
Energy and Utilities	£32,078,900	£85,186,800	£57,201,840	£72,737,880	78.32%
Other Code Measures	£10,252,100	£31,504,800	£31,393,920	£31,454,520	206.22%
TOTALS	£445,354,000	£520,948,000	£492,700,000	£508,123,000	10.63%

Cost plan exclusions:

1) Professional fees
2) Land acquisition costs
3) S106 & S278 costs
4) Service diversions
5) Protection of existing structures
6) VAT
7) Inflation beyond 1Q 2008
8) Costs arising from archaeological investigations
9) Demolition of existing structures on the site
10) Public consultation costs
11) Legal fees
12) Finance charges
13) Sales and marketing costs
14) Compensation costs
15) Amounts payable for continuing maintenance & commuted sums
16) Life cycle & maintenance costs including to beach and river defences
17) Land remediation costs

Table C6 - Cost summary

Costs

Figure C49 indicates the proportional cost breakdown. The three LifE options indicate the relative change in the energy and utilities cost of the total.

Option 2 was found to provide the most cost effective energy solution to achieving zero carbon, £28m less than option 1. This was on the basis that after using biomass to produce the majority of the energy needed the remaining electricity would be provided by large wind turbines instead of solar PVs.

Option 3, which did not use biomass and PVs, was found to be £16m more expensive than option 2. These results show that the wind power and heat exchange systems are more cost effective than solar PVs.

Integrated Approach

The cost of developing the site to Code Level 6 and combining ecological flood mitigation in a non-integrated way was assessed to be more expensive than with the LifE integrated approach, see figure C50. The integrated approach was found to save upto £46.4m, 8.6% of the total development cost.

This assessment was made on the basis that without establishing an integrated plan large wind turbines would be untenable, and the area of land required for parking, recreation and SUDS would be greater unless land-uses where combined.

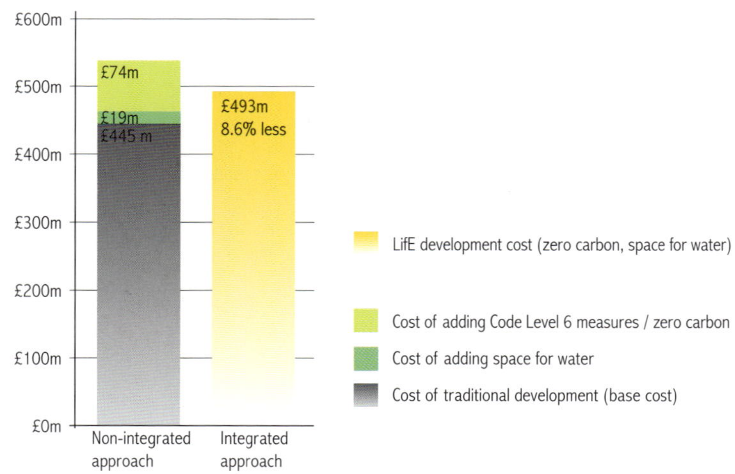

Figure C49 - Cost between integrated and non-integrated approach

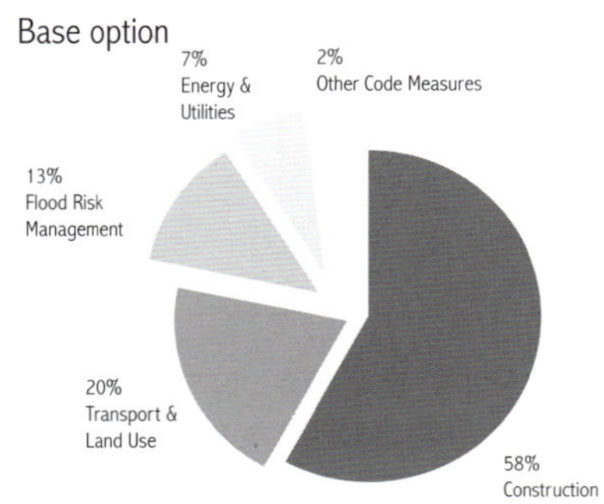

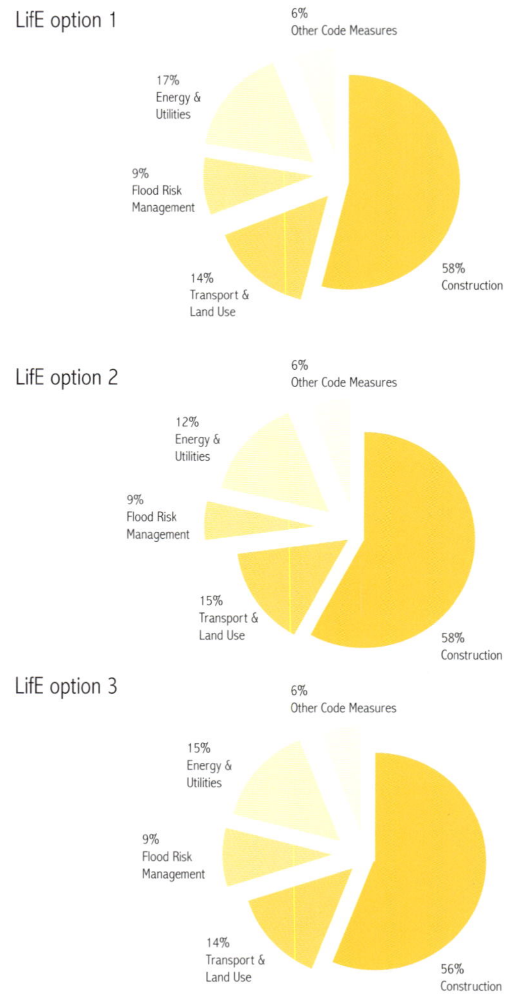

Figure C50 - Cost comparison between base and LifE option

Acceptability

The LifE project proposals were assessed against related planning policy, relevant building regulations, the Code for Sustainable Homes and a number of other guidance documents.

In general the proposals for site 3 were found to be in accordance with, if not exceeding, policy, particularly planning policy. The integration of amenity with biodiversity, energy generation and water management would be well supported in policy.

Tidal floodplain restoration and other flood storage schemes need considerable land transfer or sale. This would require landowner and community agreement, which can be difficult to obtain. In the lower Arun conflicting interests in flood risk management and designated uses, such as the SSSI at Climping Beach, could arise. The benefit of habitat enhancement and restoration of natural processes may be deemed to outway the losses.

The development of the Harbour/Westbank site would be required to pass the exception test, and would require the acceptance by the planning authority that the land once raised out of the floodplain is no longer Flood Zone 3b.

Installation of large wind turbines may be opposed on grounds of visual pollution, noise, vibration and risk to bird life.

Land-based turbines are typically cheaper than off-shore turbines due to the scale of the foundations required and the cost of providing underwater power cables. Therefore to meet renewable energy targets wind power within developments needs to be considered seriously.

Appropriateness

Many of the existing riverside properties in Littlehampton are built against existing defences. Maintaining a consistent defence to this area is difficult to ensure without the creation of a completely new defensive structure, such as sheet piling. This would be expensive and further reduce the width of the river channel. Alleviating pressure on existing defences by diverting peak tide flows to reduce flood levels could help to maintain the defence's active life.

Localised defences around key infrastructure may only provide a short-term solution to flood risk if sea levels increase more rapidly than predicted. Therefore facilities might need to be relocated in the future.

Adaptability

The effects of climate change on site 3 come from sea level rise, increased storminess, increased intensity of peak rainfall and increased river flows. The site will become more susceptible to extreme events. This is likely to be the case in most low-lying coastal sites around the UK. Therefore the need for an adaptable solution was critical.

The flood defences provide protection to a certain level but the standard will reduce with time and the effects of climate change. A residual risk from overtopping or failure of the defences would remain as would the need for maintenance. Therefore reducing dependence on flood defences was an important part of the conceptual solution for the lower Arun. Ultimately if sea levels rise faster than expected there will be a point where the longevity of the whole area will be in question; and consideration of that possibility now to prevent unsustainable development is vital.

The LifE proposals provided a possible approach to short-term development as part of a plan for long-term adaptation.

Flood Risk Assessment

1	**Development description and location**	**The lower Arun at Littlehampton**
1a	What type of development is proposed and where will it be located?	Refer to Masterplan drawings
1b	What is its vulnerability classification?	Combination of more and less vulnerable
1c	Is the proposed development consistent with the Local Development Documents?	Yes
1d	Please provide evidence that the Sequential Test and where necessary the Exception Test have been applied in the selection of this site for this development type?	N/A. The three study sites were selected during Stage 1 of the project from a number of real sites submitted by local government from around the country. The three sites were chosen on the basis of providing a range of detailed issues to test the 'LifE' strategy and providing a number of flood risk issues that would be most representative of issues likely on other sites.
2	**Definition of the flood hazard**	
2a	What sources of flooding could affect the site? (see annex C PPS25).	Flooding from the Sea Flooding from Rivers Flooding from Land Flooding from Groundwater Flooding from Sewers
2b	For each identified source, describe how flooding would occur, with reference to any historic records wherever these are available.	Water levels in the study area are dominated by coastal conditions (high tides are constrained by the existing river channel training walls at the mouth of the River Arun, risk of tidal flooding during storms, risk of failure of the river channel training walls, risk of a breach in the existing shingle ridge adjacent to the mouth of the River Arun). Fluvial flows along the River Arun are typically sluggish in summer and prone to flash flooding in winter. Fluvial flows along the River Arun could be significant during a severe storm event.
2c	What are the existing surface water drainage arrangements for the site?	Unknown. Not investigated as part of this study. The scope of work for this study did not involve the development of full FRAs for each of the three sites.
3	**Probability**	
3a	Which flood zone is the site within?	Flood Zones 1, 2 & 3
3b	If there is a Strategic Flood Risk Assessment covering this site, what does it show?	No SFRA available for this site
3c	What is the probability of the site flooding, taking account of the contents of the SFRA and of any further site-specific assessment?	Functional floodplain
3d	What are the existing rates and volumes of run-off generated by the site?	Unknown. Not investigated as part of this study. The scope of work for this study did not involve the development of full FRAs for each of the three sites.
4	**Climate change**	
4a	How is flood risk at the site likely to be affected by climate change?	Sea level rise will increase the risk of tidal flooding to the site and increase the probability of failure of the river channel training walls and increase the probability of a breach in the shingle ridge
5	**Detailed development proposals**	
5a	Please provide details of the development layout, referring to the relevant drawings.	Refer to Masterplan drawings
5b	Where appropriate, demonstrate how land uses most sensitive to flood damage have been placed in areas within the site that are at least risk of flooding.	Refer to Masterplan drawings
6	**Flood risk management measures**	
6a	How will the site be protected from flooding, including the potential impacts of climate change, over the development's lifetime?	Maintain the river channel training walls, shingle ridge and provision of Flood Control Areas (tidal lagoons and inter-tidal habitats) to reduce the risk of flooding to Littlehampton and the proposed development area.
7	**Off-site impacts**	
7a	How will you ensure that your proposed development and the measures to protect your site from flooding will not increase flood risk elsewhere?	See above
7b	How will you prevent run-off from the completed development causing an impact elsewhere?	Provision of SUDS linked to Flood Control Areas (tidal lagoons and inter-tidal habitats)
8	**Residual risks**	
8a	What flood-related risks will remain after you have implemented the measures to protect the site from flooding?	Future sea level rise will limit the operational life of the development, following which it will be necessary to relocate residential and non-residential properties and infrastructure inland.
8b	How, and by whom, will these risks be managed over the lifetime of the development?	Environment Agency (beach management plan), Littlehampton Harbour Authority (river channel training walls) and TBC (development)

Table C7 - Flood Risk Assessment (FRA)

The proposals for the lower Arun at Littlehampton showed that the flood risk could be managed with a combination of non-defensive and defensive measures to reduce the level of risk. These measures included:

- avoiding risk by locating development in the areas of lowest probability of flooding and locating less vulnerable uses on the lower floors;
- reducing the level of flood water by providing flood storage, raising and improving river defences, raising areas of land, realigning the coastal defence to create a more natural bay shape, providing rainwater attenuation and preparing a long-term strategy for the lower Arun;
- mitigating the effects of flooding by use of resilient building construction and providing dedicated channels to return floodwater to the river channel after a flood.

The flood storage areas were shown to provide environmental gain, recreational benefit and opportunity for renewable energy generation in the form of tidal and large-scale wind power. However, the scale of the flood storage areas matched the scale of the flood risk, and the flood storage areas required were found to be significantly larger than the site area of the development.

The range of natural resources suitable for generating energy enabled several options to be considered to provide all of the energy needs of the development. This allowed more cost effective technologies to be considered and for the innovative integration of tidal power generation with flood storage controls. zero carbon development of sites in the lower Arun catchment, for approximately 2000 homes, was found to be technically feasible. On the basis of available resources a larger number of zero carbon homes could be built in this area. The energy sources were thought to be able to generate sufficient power to provide a surplus to the needs of the development alone. This surplus could be used to supply existing dwellings.

Safe redevelopment of the harbour site was considered feasible, with land raising, new defences and resilient building construction. Buildings located behind defences were found to be at risk of flooding from overtopping or failure of defences. Therefore location of appropriate uses at ground floor and resilient construction were required. Elevation of main roads and emergency services was required to provide safe access. Level access into buildings was found to be possible through spatial planning around access routes, however, changes in level required more space for ramps than conventional design. These ramps and changes in level could be integrated with recreational uses and rainwater attenuation to provide attractive neighbourhoods. The longevity of the use of this site was thought to become more questionable with sea level rise over 3 m to 4 m. However, with sea level rise of this magnitude the economic sustainability of many existing coastal settlements would need to be considered.

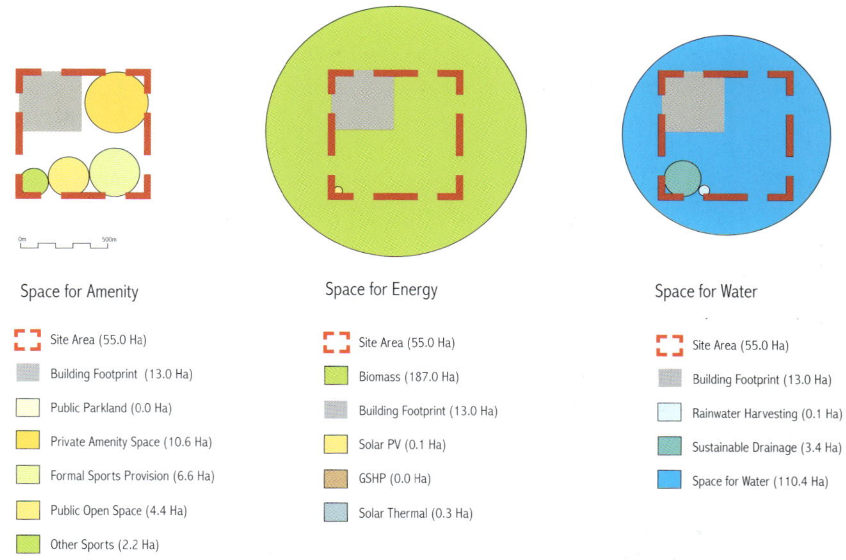

Figure C51 - Total space demands for the Climping and Harbour Sites at Littlehampton.

Outdoor play requirements for the residents of the new development were exceeded through provision on multi-use land within the development, along the riverbanks and in the landscape around the flood storage areas. This landscape could also provide space for crop production and grazing.

Figure C51 illustrates the spatial requirements for the sites and the various provisions for amenity space and flood mitigation.

> **The lower catchment**
>
> The predominant flood risk from the sea and tidal river led to the development of the key message:
>
> ***'let tides go'***
>
> This means that tidal flooding is to be avoided where possible, to direct floodwater around sites.

APPENDIX D
Assumptions

Zero Carbon Definition

There is currently much debate regarding the definition of zero carbon for the built environment. The government's current published definition for dwellings can be found in the Code for Sustainable Homes[3] (April 2008 version). This requires that net CO_2 emissions from the dwelling over the course of a year from all energy uses in the home should be zero, and that this must be achieved solely by the use of energy generated on site.

At the time of analysis for LifE, the Code definition of zero carbon had just been changed (October 2007 update of the Code), and there was much uncertainty regarding what a final workable definition of zero carbon might be. Therefore, for the purposes of the LifE analysis, the following definition was adopted:

All building energy demands will be met by the use of low or zero carbon energy technologies, with the aim that the demands will be met solely by on-site energy generation. The means of energy generation should draw on the resources of the development site where possible. Where it is not physically possible to meet all demands on site, use of off-site electricity generation will be allowed as long as it is accredited renewable generation capacity outside of the renewables obligation (RO). However it must first be demonstrated that all viable on-site generation opportunities have been exhausted.

The first part of this definition is subtly different from the Code definition in that it doesn't take advantage of the current carbon factor advantage for on-site generated electricity (as this is a much debated point). The second part acknowledges the fact that for some sites it may not actually be possible to generate enough low or zero carbon energy on site in order to be 'Zero Carbon'.

In May 2008, the UK Green Building Council published a report entitled 'The Definition of Zero Carbon'[51] which looked at some of the reported problems with the Code definition and established a task group of industry experts to assess the need for change. The report found that, depending on how rigorously carbon was counted by the calculation methodology, anywhere between 10% & 80% of new homes would be physically unable to meet the governments definition of zero carbon (regardless of capital expenditure). The task group concluded that, in order to maintain the ambition of the original government policy statement, a revised definition of 'Zero Carbon' that allows the use of off-site solutions in certain circumstances would be required.

Energy Assumptions

Energy demand for domestic buildings

Residential Buildings

Space Heating: Assume PassivHaus Standard of 15 kWh/m²/yr

Space Cooling: Assume none.

Hot Water Heating: From SAP2005[52] check for a previous scheme, reduced by 10% to take into account water efficient fittings and appliances.

Electricity (excluding cooling) from BREDEM-12[53] calculation and SAP2005[52] results, reduced by 10% to take into account more efficient white goods and change in behaviour

Energy demand for non-domestic buildings

Office

Space Heating: Good practice heating & hot water use, air conditioned office = 97 kWh/m² - Split 80:20 heating:hot water (ECG 19, 2000[54]), assume 80% efficient boilers to convert delivered energy to energy demand, assume 60% reduction in demand due to additional energy efficiency measures.

Space Cooling: Good practice cooling, air conditioned office (ECG 19, 2000[54]), assumed Coefficient of Performance (COP) = 3.

Hot Water Heating: Good practice heating & hot water use, air conditioned office = 97 kWh/m² - Split 80:20 heating:hot water (ECG 19, 2000[54]), assume 80% efficient boilers to convert delivered energy to energy demand, assume 10% reduction in demand due to low water use appliances etc.

Electricity (excluding cooling) Good practice electricity use (fans pumps & controls, lighting, office equipment, catering, other electric appliances, computer room), air conditioned office (ECG 19, 2000). Assume 10% reduction in demand due to additional energy efficiency measures and change in behaviour.

Retail

Space Heating: Good practice, non-food store, heating & hot water = 82 kWh/m² - Split 90:10 heating:hot water (CIBSE Guide F[55]), assume 80% efficient boilers to convert delivered energy to energy demand, assume 60% reduction in demand due to additional energy efficiency measures.

Space Cooling: Assumed equal to 10% of total electric load (extrapolated from stats in GIL143[56]).

Hot Water Heating: Good practice, non-food store, heating & hot water = 82 kWh/m² - Split 90:10 heating:hot water (CIBSE Guide F[55]), assume 80% efficient boilers to convert delivered energy to energy demand, assume 10% reduction in demand due to low water use appliances etc.

Electricity (excluding cooling): Good practice, non-food store, total electricity minus cooling electricity (CIBSE Guide F[55]), assume 20% reduction in demand due to additional energy efficiency measures and change in behaviour.

Industrial

Space Heating: New 'Storage and distribution' building (ECG 18, EEBPP, 1993[57]), heating & hot water = 80 kWh/m² - split 70:30 heating: hot water, assume 80% efficient boiler to convert delivered energy to energy demand, assume 60% reduction in demand due to additional energy efficiency measures.

Space Cooling: None assumed.

Hot Water Heating: New 'Storage and distribution' building (ECG 18, EEBPP, 1993[57]), heating & hot water = 80 kWh/m² - split 70:30 heating: hot water, assume 80% efficient boiler to convert delivered energy to energy demand, assume 10% reduction in demand due to low water use appliances etc.

Electricity (excluding cooling): New 'Storage and distribution' building (ECG 18, EEBPP, 1993[57]).

Educational

Space Heating: School C (BSRIA Delta magazine article, June 2007[58]) 88 kWh/m²/yr - assumed split 60:40 space heating: hot water, assume 90% efficient boiler to convert delivered energy to energy demand, assume 40% reduction in demand due to additional energy efficiency measures.

Space Cooling: School C (BSRIA Delta magazine article, June 2007[58]), COP 3 assumed to turn electricity load to cooling load.

Hot Water Heating: School C (BSRIA Delta magazine article, June 2007[58]), 88 kWh/m²/yr - assumed split 70:30 space heating: hot water, assume 90% efficient boiler to convert delivered energy to energy demand, assume 10% reduction in demand due to low water use appliances etc.

Electricity (excluding cooling): School C (BSRIA Delta magazine article, June 2007[58]), assume 10% reduction in demand due to additional energy efficiency measures.

Cost Assumptions - Site 1, Hackbridge

Base Cost Plan

Construction

- 100% traditional residential construction
- All residential construction to meet Code for Sustainable Homes Level 3
- Traditional build retail, commercial and industrial units
- All retail, commercial and industrial units to meet BREEAM excellent
- Traditional build education, health, community and car club facilities
- All education, health, community and car club units to meet BREEAM excellent

Transport

- Street level parking including paths and roads
- Standard on-site road construction
- 10% of undeveloped site area to be hard landscaped
- 30% of undeveloped site area to be soft landscaped
- Soft landscaping for private amenity space such as private gardens
- Private amenity space based upon 5 m^2 per person

Site Services Infrastructure

- Conventional site-wide electrical infrastructure
- Conventional site-wide water infrastructure
- Conventional site-wide gas infrastructure
- Conventional site-wide waste water and surface water drainage infrastructure
- Conventional site-wide telecommunications infrastructure
- External lighting to paths and parking areas

Site Clearance, Land Remediation & Use

- Site clearance includes for slab break up, levelling and compacting of site.
- Allowance for multi-use play areas
- Allowance for sports pitches
- Allowance for allotments

Flood Risk Management

- Compensatory excavation to levels of 0.1 m and 0.2 m
- Allowance for land raising

Off-Site Works

- Allowance for the incorporation of civic art

LifE Option 1 Cost Plan

Construction

- 60% traditional residential construction
- 30% flood resistant (dry-proof) residential construction
- 10% elevated residential construction
- All residential to meet Code for Sustainable Homes Level 6
- Traditional build retail
- Wet-proof construction to commercial and industrial units
- All retail, commercial and industrial units to meet BREEAM excellent
- Traditional build education, health, community and car club facilities
- All education, health, community and car club units to meet BREEAM excellent
- Car club to include 200 m^2 of office space and 55 car parking spaces

Transport

- Reduced street level car parking
- Standard on-site road construction
- 18% of undeveloped site area to be hard landscaped
- 55% of undeveloped site area to be soft landscaped
- Soft landscaping for private amenity space such as private gardens
- Private amenity space based upon 5m^2 per person

Site Services Infrastructure

- 430 kW Photovoltaic Cells
- 9 kW Micro Hydro Power
- 200 kWe Biomass Combined Heat and Power
- 2000 kW Biomass Boiler
- 550 kW Ground Source Heat Pump
- 710 kW Solar Thermal Hot Water
- Conventional site-wide electrical infrastructure
- Conventional site-wide water infrastructure
- No site-wide gas infrastructure
- Conventional site-wide waste water and surface water drainage infrastructure
- Conventional site-wide telecommunications infrastructure
- External lighting to paths, road, parking, hard and soft landscaping

Site Clearance, Land Remediation & Use

- Site clearance includes for slab break up, levelling and compacting of site.
- Allowance for Bio Crop production for Biomass CHP and boiler
- Allowance for multi-use play areas
- Allowance for sports pitches
- Allowance for allotments

Flood Risk Management

- Allowance for 1.5 m (average) deep dry pond
- Allowance for SUDS in the forms of swales, rain gardens and dry ponds
- Re-naturalisation excavation to level of 0.5 m

Off-Site Works

- Allowance for the incorporation of civic art

Cyril Sweett - LifE Project Cost Plan Assumptions

Cost Assumptions - Site 2, Peterborough

Base Cost Plan

Construction

- 99% traditional residential construction
- 1% floating residential construction
- All residential construction to meet Code for Sustainable Homes Level 3
- Traditional build retail, commercial and industrial units
- All retail, commercial and industrial units to meet BREEAM excellent
- Traditional build education, health, community and car club facilities
- All education, health, community and car club units to meet BREEAM excellent

Transport

- Street level parking including paths and roads
- Standard on-site road construction
- 12% of undeveloped site area to be hard landscaped
- 36% of undeveloped site area to be soft landscaped
- Soft landscaping for private amenity space such as private gardens
- Private amenity space based upon 5 m^2 per person

Site Services Infrastructure

- Conventional site-wide electrical infrastructure
- Conventional site-wide water infrastructure
- Conventional site-wide gas infrastructure
- Conventional site-wide drainage infrastructure
- Conventional site-wide telecommunications infrastructure
- External lighting to paths and parking areas

Site Clearance, Land Remediation & Use

- Site clearance includes for slab break-up, levelling and compacting of site.
- Allowance for multi-use play areas
- Allowance for sports pitches
- Allowance for allotments

Flood Management

- Concrete bank reinforcement to 850 m of development
- Allowance for area of land raising by 0.2 m

Off-Site Works

- Allowance for the incorporation of civic art

LifE Options Cost Plan

Construction

- 43% traditional residential construction
- 31% flood resistant (dry-proof) residential construction
- 11% elevated residential construction
- 12% flood resilient (wet-proof) residential construction
- 2% amphibious residential construction
- 1% floating construction
- All residential to meet Code for Sustainable Homes Level 6
- Traditional build retail
- Wet-proof construction to commercial and industrial units
- All retail, commercial and industrial units to meet BREEAM excellent
- Traditional build education, health, community and car club facilities
- All education, health, community and car club units to meet BREEAM excellent
- Car club to include 200 m^2 of office space and 100 car parking spaces

Transport

- Reduced street level car parking
- Standard on-site road construction
- Broadwalk allowance for pedestrian paths
- 16% of undeveloped site area to be hard landscaped
- 56% of undeveloped site area to be soft landscaped
- Soft landscaping for private amenity space such as private gardens
- Private amenity space based upon 5 m^2 per person
- Pontoon construction for ferry ports

Site Services Infrastructure

- Conventional site-wide electrical infrastructure
- Conventional site water infrastructure
- No site-wide gas infrastructure
- Conventional site waste water infrastructure
- Reduced provision for site surface water drainage infrastructure
- Conventional site-wide telecommunications infrastructure
- External lighting to paths, road, parking, hard and soft landscaping

Site Clearance, Land Remediation & Use

- Site clearance includes for slab break-up, levelling and compacting of site
- Allowance for area of anaerobic digestion
- Allowance for Bio Crop production for Biomass CHP and boiler
- Allowance for multi-use play areas
- Allowance for sports pitches
- Allowance for allotments

Flood Risk Management

- Green roofs to residential construction in high risk areas only
- Allowance for SUDS underground storage
- Allowance for SUDS in the forms of swales, rain gardens and dry ponds

Off-Site Works

- Allowance for the incorporation of civic art

Life Option 1 Cost Plan

Site Services Infrastructure

- 10417 kW Photovoltaic Cells
- 720 kW Wind Power
- 500 kWe Biomass Combined Heat and Power
- 5000 kW Biomass Boiler
- 1500 kW ATES Heat Pump
- 1895 kW Solar Thermal Hot Water

Life Option 2 Cost Plan

Site Services Infrastructure

- 8360 kW Photovoltaic Cells
- 1800 kW Wind Power
- 500 kWe Biomass Combined Heat and Power
- 5000 kW Biomass Boiler
- 1500 kW ATES Heat Pump
- 1895 kW Solar Thermal Hot Water

Cyril Sweett - LifE Project Cost Plan Assumptions

Cost Assumptions - Site 3, Littlehampton

Base Cost Plan

Construction

- 100% traditional residential construction in the Harbour & Climping areas
- All residential construction to meet Code for Sustainable Homes Level 3
- Traditional build retail, commercial and industrial units
- All retail, commercial and industrial units to meet BREEAM excellent
- Traditional build education, health, community and car club facilities
- All education, health, community and car club units to meet BREEAM excellent

Transport

- Street level parking including paths and roads
- Standard on-site road construction
- 15% of site area to be hard landscaped
- 44% of site area to be soft landscaped
- Soft landscaping for private amenity space such as private gardens
- Private amenity space based upon 5 m² per person

Site Services Infrastructure

- Conventional site-wide electrical infrastructure
- Conventional site-wide water infrastructure
- Conventional site-wide gas infrastructure
- Conventional site-wide waste water and surface water drainage infrastructure
- Conventional site-wide telecommunications infrastructure
- External lighting to paths and parking areas

Site Clearance, Land Remediation & Use

- Site clearance includes for slab break-up, levelling and compacting of site.
- Allowance for multi-use play areas
- Allowance for sports pitches
- Allowance for allotments

Flood Risk Management

- Allowance for raising of river bank by 3 m over 2000 m
- Mix of sheet pile and gabion reinforcement to length of development
- Allowance for training walls from beach
- Allowance for River Arun defence improvements
- Allowance for land raising by up to 3 m to Harbour area
- Allowance for sewage plant protective dyke
- Allowance for other River Arun defence improvements

Off-Site Works

- Allowance for the incorporation of civic art

LifE Options Cost Plan

Construction

- 50% flood resilient (wet-proof) residential construction in Harbour area
- 40% elevated residential construction in Harbour area
- 10% floating residential construction in Harbour area
- 80% traditional residential construction in Climping area
- 20% flood resistant (dry-proof) residential construction in Climping area
- All residential to meet Code for Sustainable Homes Level 6
- Wet-proof construction to retail, commercial and industrial units
- All retail, commercial and industrial units to meet BREEAM excellent
- Traditional build education, health, community and car club facilities
- All education, health, community and car club units to meet BREEAM excellent
- Car club to include 200 m² of office space and 150 car parking spaces

Transport

- Reduced street level car parking
- Standard on-site road construction
- 18% of site area to be hard landscaped
- 55% of site area to be soft landscaped
- Private amenity space based upon 5 m² per person
- Pontoon construction for floating dwellings

Site Services Infrastructure

- Conventional site-wide electrical infrastructure
- Conventional site-wide water infrastructure
- No site-wide gas infrastructure
- Conventional site-wide waste water drainage infrastructure
- Reduced provision of surface water drainage infrastructure
- Conventional site-wide telecommunications infrastructure
- External lighting to paths, road, parking, hard and soft landscaping

Site Clearance, Land Remediation & Use

- Site clearance includes for slab break-up, levelling and compacting of site
- Allowance for Bio Crop production for Biomass CHP and boiler
- Allowance for wetland regeneration
- Allowance for multi-use play areas
- Allowance for sports pitches
- Allowance for allotments

Flood Risk Management

- Allowance for retention lakes
- Green roofs to residential construction in high risk areas only
- Allowance for relief channels
- Allowance for SUDS
- Allowance for raising of river bank by 3 m over 2000 m
- Mix of sheet pile and gabion reinforcement to length of development
- Lagoon bank reinforcement
- Allowance for training walls from beach
- Allowance for reduced River Arun defence improvements
- Allowance for land raising by average 1 m to Harbour area
- Allowance for sewage plant protective dyke
- Allowance for other River Arun defence improvements

Off-Site Works

- Allowance for the incorporation of civic art
- Allowance for floating gardens within lagoon

Life Option 1 Cost Plan - Site Services

- 2000 kW Wind Power
- 5033 kW Photovoltaic Cells
- 491 kW Tidal Power
- 500 kWe Biomass Combined Heat and Power
- 5100 kW Biomass Boiler
- 1500 kW ATES Heat Pump
- 1771 kW Solar Thermal Hot Water

Life Option 2 Cost Plan - Site Services

- 4000 kW Wind Power
- 33 kW Photovoltaic Cells
- 491 kW Tidal Power
- 500 kWe Biomass Combined Heat and Power
- 5100 kW Biomass Boiler
- 1500 kW ATES Heat Pump
- 1771 kW Solar Thermal Hot Water

Life Option 3 Cost Plan - Site Services

- 8000 kW Wind Power
- 491 kW Tidal Power
- 4000 kW Ground Source Heat Pump
- 1500 kW ATES Heat Pump
- 4678 kW Solar Thermal Hot Water

Cyril Sweett - LifE Project Cost Plan Assumptions

Addendum

Improvements to flood risk and climate change modelling may result in changes to flood mapping. This highlights the importance of working with the most up-to-date information and adopting a precautionary and flexible approach, as advocated by the LifE Project.

At the time of writing the LifE report the SFRA's for both the River Wandle and River Nene were in draft format.

Site 1 - Updated Flood Maps

The joint SFRA for the London Boroughs of Wandsworth, Merton, Sutton and Croydon was published in December 2008. Figure A43 indicates the extent of the flood zones in relation to the Hackbridge sites. This map shows the 1 in 100 year flood event and Flood Zone 3a to be more extensive than previously expected. For residential development within Flood Zone 3a to be accepted a sequential test would need to be carried out to demonstrate that alternative sites could not be found and the exception test passed to demonstrate that the development is safe.

The London Borough of Sutton published its PPS25 Sequential Test in 2008[57]. This established that the case for development on a range of sites around the borough. This document establishes the requirement to use the Hackbridge sites for development, with a large residential component.

'The London Plan target for LB Sutton is 345 additional residential dwellings per year, with the intention of the Council to exceed this target. In order for these targets to be met over the next 10 years

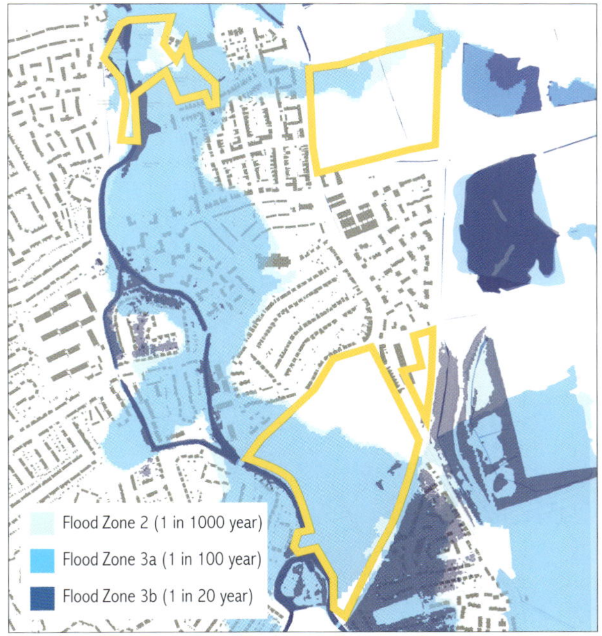

Figure D1 - Site 1 flood risk map current (2009) based on SFRA level 1

all available sites will need to be exploited, including all available development sites located within Flood Zone 1.'[59]

This new information may require some residential buildings suggested to be changed to less vulnerable uses, or elevated above the flood level and loss of floodplain compensated for. Areas of SUDS in Flood Zone 3 would need to be elevated above ground level.

Site 2 - Updated Flood Maps

The level one SFRA for Peterborough was completed in 2008 for Peterborough City Council. Figure D2 shows the extent of Flood Zone 3

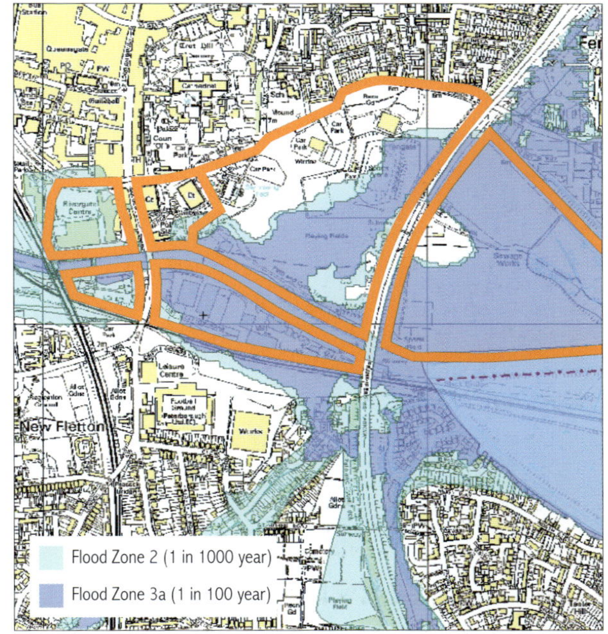

Figure D2 - Site 2 flood risk map current (2009) from the Environment Agency

to be similar to that shown on the Environment Agency website. However, this does not consider tidal influences that may occur as a result of sea level rise and the Environment Agency expect this to change following current remodelling work.

This further modelling work, yet to be published, may put into question the development of the Fengate site, particularly if it does not adopt an adaptable approach.

The measures proposed in the LifE project may be possible to be adapted in response to the new results and/or form a leading pilot project to explore adaptability to climate change.

References and Glossary >

References

1. The Environment Agency (2008) (online) www.environment-agency.gov.uk/subjects/flood/
2. Scottish Executive, 2008 Climate Change Flooding Occurrences Review (online) www.sepa.org.uk/pdf/publications/leaflets/flood/floodline_leaflet_05.pdf
3. CLG (2008) Code for Sustainable Homes: Technical Guide: Communities and Local Government
4. Lifetime Homes (2008) (online) www.lifetimehomes.org.uk
5. NPFA (1992) The Six Acre Standard: Minimum Standards for Outdoor Playing Space
6. Secured by Design (2008) (online) www.securedbydesign.com
7. EU (2003) Water Framework Directive
8. Defra (2008) Making Space for Water (online): www.defra.gov.uk
9. CLG (2006) Planning Policy Statement 25: Development and Flood Risk: Communities and Local Government www.defra.gov.uk
10. CLG (2008) PPS25: Practice Guide: Communities and Local Government
11. Sir Michael Pitt (2008) Pitt review: Learning Lessons from the 2007 floods: Cabinet Office
12. UN (2004) World Population to 2300: United Nations
13. ODPM (2007) Sustainable Communities: Building for the Future: Office of the Deputy Prime Minister (now CLG)
14. CLG (2008) Sustainable Communities (online) www.communities.gov.uk
15. OCC (2005) Stern Review on the Economics of Climate Change: OCC Publications
16. IPCC (2007) Climate Change 2007: Synthesis Report: IPCC
17. Defra (2006) The Flood and Coastal Defence Appraisal Guidance FCDPAG3, Supplementary Note to Operating Authorities – Climate Change Impacts: Department for Food and Rural Affairs
18. UNFCCC (2005) Kyoto Protocol (online): www.kyotoprotocol.com
19. Defra: UK Climate Impacts Programme (2002) The UKCIP02 Scientific Report: Climate Change Scenarios for the United Kingdom, Tyndall Centre
20. Baca Architects with BRE (2009) LifE Handbook: IHS BRE Press
21. English Nature (2003) Accessible Natural Green Space Standards in Towns and Cities: A Review and Toolkit for Implementation: English Nature
22. Building Regulations Approved Documents (Part B, L and M): The Stationary Office
23. EA (2007) Thames Region Catchment Flood Management Plan, summary document, Consultation: Environment Agency
24. EA (2008) Catchment Flood Management Plan, River Nene, Draft (unpublished)
25. EA (2006) Arun and Western Streams Catchment Flood Management Plan (CFMP) Consultation Draft Plan: Environment Agency
26. Europa (2007) European Birds and Habitats Directives
27. Housing Corporation (2007) Housing Quality Indicators Version 4: Housing Corporation
28. CLG (2003) Planning Policy Statement 1: Delivering Sustainable Development: Communities and Local Government
29. CLG (2006) Planning Policy Statement 3: Housing: Communities and Local Government
30. CLG (2005) Planning Policy Statement 9: Biodiversity and Geological Conservation: Communities and Local Government
31. CLG (2004) Planning Policy Statement 22: Renewable Energy: Communities and Local Government
32. CLG (2002) Planning Policy Guidance 17: Planning for Open Space, Sport and Recreation: Communities and Local Government
33. The Green Flag Award (2008) Guidance Manual (online). www.greenflagaward.org.uk/manual/
34. Scott Wilson (2008) Strategic Flood Risk Assessment, Level 1, River Wandle: London Borough of Sutton
35. Royal Haskoning (2008) Strategic Flood Risk Assessment Update - Level 1 Report: Peterborough City Council
36. Ciria (2008) Sustainable Drainage Systems: Promoting Good Practice: www.ciria.org/suds
37. Pam Bowker (2007) Flood resistance and resilience solutions: an R&D scoping study, Defra
38. Association of British Insurers (2004) turning back the tide: The case for sustaining investment in flood management and defences, ABI
39. Association of British Insurers (2007) Summer floods 2007: Learning the lessons
40. London (2008) (online). www.london.gov.uk
41. Llewelyn, Davies, Roger Tym and Partners with Campbell Reith and Steer Davies Gleave (2004) Peterborough Growth Area Study: Final Report
42. Wikipedia (2008) (online): www.wikipedia.org
43. English Heritage (2008) (online): www.english-heritage.org.uk
44. Defra (2008) UKCIP 1971-2000 (Trends20_totalprecip_71to00.pdf) (online): www.ukcip.org.uk
45. South East England Regional Assembly (2006) draft South East Plan (online): www.southeast-ra.gov.uk
46. Donaldsons/BDP (2004) The Littlehampton Vision, Final report
47. HR Wallingford (2002) River Arun to Pagham Coastal Defence Strategy Study, Executive Summary, Report EX 4466. Report produced for Arun District Council and the Environment Agency.
48. Halcrow (2006) Beachy Head to Selsey Bill Shoreline Management Plan: Report produced for the South Downs Coastal Group.
49. Royal Haskoning (2007) Tidal River Arun, Initial Geomorphological Assessment, Draft Report 2, Royal Haskoning for Environment Agency, West Sussex, 23 April 2007
50. DJ Leggett, N Cooper & R Harvey (2004) Coastal and estuarine managed realignment – design issues: CIRIA Report C628
51. UK Green Building Council (2008) Definition of Zero Carbon
52. BRE (2005) Standard Assessment Procedure (SAP)
53. BRE (2002) BREDEM-12: The BRE Domestic Energy Model (2001 update): BRE Press
54. Carbon Trust (2000) Energy Efficiency - Best Practice Programme (EEBPP) - Energy Consumption Guide 019: Energy Use in Offices
55. Chartered Institution of Building Services Engineers (2004) CIBSE Guide F: Energy efficiency in buildings
56. Carbon Trust (2004) Energy Saving Fact Sheet - Retail. GIL 143
57. Carbon Trust (1993) Energy Efficiency - Best Practice Programme (EEBPP) - Energy Consumption Guide 018
58. BSRIA (2007) Delta magazine article
59. London Borough of Sutton (2008) PPS25 Sequential Test, Based on Joint Strategic Flood Risk Assessment, Evidence Base

Also referred to but not cited

CIRIA (2005) Designing for exceedence (online). www.ciwem.org/groups/wapug/S2005_DigmanShaffer.pdf

BRE (2002) Sustainability Checklist for Developments: BRE Press

CLG (2006) Code for Sustainable Homes – A step change in sustainable home building practice

JBA (2004) Extreme Sea Levels Kent, Sussex, Hampshire and Isle of Wight

Flows report WP3Cvii-1, Sustainable Drainage Systems for new homes – best practice guidance, June 2006

CLG (2007) Improving the Flood Performance of New Buildings, Flood Resilient Construction

Micro Hydro (online) www.hydrogeneration.co.uk

Civil Engineering (2002) Old Watermills

Cambridgeshire County Council (1991) Cambridgeshire Landscape Guidelines

The Landscape Institute and Institute of Environmental Management and Assessment (2002) Guidelines for Landscape and Visual Impact Assessment, (Second Edition): Spon Press

Office of Science and Technology (2004) Executive Summary – Future Flooding, A Foresight Report

Anderson, S.H (2001) Barrages and Impoundments, a preliminary study of the potential for sustainable net energy capture obtainable by storage for tidally amplified release ('ecostar' scoping model) (online). www.marinet.org.uk

Image Credits

1. Baca Architects (2009) LifE Handbook: This is not the only solution: IHS BRE Press
2. Baca Architects (2006) Working with water: The LifE Approach
3. Baca Architects (2009) Consideration of the wider context is important to avoid adversely affecting neighbours.
4. Baca Architects (2009) Diagram of an integrated spatial plan
5. Baca Architects (2009) Spatial Planning Process and the stages influenced by the LifE project
6. CLG (2008) The Communities plan growth areas (online): www.communities.gov.uk © Crown copyright 2008
7. Baca Architects (2007) The LifE project process diagram
8. Baca Architects (2009) Site 1 planning based on walking distances
9. Baca Architects (2009) The LifE study locations
10. Baca Architects (2009) The River Wandle at Hackbridge
11. EA (2008) Flood Risk Map for site 1 (online): www.environment-agency.gov.uk © Environment Agency copyright 2008. All rights reserved
12. Baca Architects (2009) The River Nene at Peterborough
13. EA (2008) Flood Risk Map for site 2 (online): www.environment-agency.gov.uk © Environment Agency copyright 2009. All rights reserved
14. Baca Architects (2009) The River Arun at Littlehampton
15. EA (2008) Flood Risk Map for site 3 (online): www.environment-agency.gov.uk © Environment Agency copyright 2009. All rights reserved
16. Baca Architects (2009) LifE Handbook: Site 1: IHS BRE Press
17. Baca Architects (2009) Site Plan
18. Baca Architects (2009) Aerial view
19. Baca Architects (2009) LifE Handbook: Site 2: IHS BRE Press
20. Baca Architects (2009) Site Plan
21. Baca Architects (2009) Aerial view
22. Baca Architects (2009) LifE Handbook: Site 3: IHS BRE Press
23. Baca Architects (2009) Site Plan
24. Baca Architects (2009) Aerial view
25. Halcrow Group Ltd (2008) Site 1 flood risk map current: Based on draft SFRA level 1 information provided by the Environment Agency in 2008, © Environment Agency copyright 2009. All rights reserved, © Crown copyright Ordnance Survey. All rights reserved. 2009
26. Halcrow Group Ltd (2008) Site 1 flood risk map future: Based on draft SFRA level 1 information provided by the Environment Agency in 2008: © Environment Agency copyright 2009. All rights reserved
27. Baca Architects (2009) LifE Handbook: Potential effects of climate change: IHS BRE Press
28. Fulcrum consulting (2008) Energy Split Diagrams for zero carbon development
29. Fulcrum consulting (2008) Space requirements for renewable energy technologies at site 1, Hackbridge Space requirements for renewable energy technologies required to achieve zero carbon development at site 1, Hackbridge
30. Baca Architects/LDA Design (2008) Space requirements for amenity at Hackbridge
31. Cyril Sweett (2008) Cost comparison between base case and LifE option for the case studies
32. Cyril Sweett (2008) Comparison cost of un-integrated and integrated design approach.
33. Cyril Sweett (2008) Potential payback period for LifE versus traditional approach
34. Baca Architects (2009) LifE Handbook: Catchment icons: IHS BRE Press
35. Baca Architects (2009) LifE Handbook: Upper Catchment: IHS BRE Press
36. Baca Architects (2009) LifE Handbook: Middle Catchment: IHS BRE Press
37. Baca Architects (2009) LifE Handbook: Lower Catchment: IHS BRE Press
38. Baca Architects (2009) LifE Handbook: LifE Lessons: IHS BRE Press

APPENDIX A

A1. Baca Architects (2009) Location Plan
A2. Baca Architects (2008) The site areas: Based on information provided by Ordnance Survey in 2008. © Crown copyright Ordnance Survey. All rights reserved. 2009
A3. Baca Architects (2008) Key to photographs: Based on information provided by Ordnance Survey in 2008. © Crown copyright Ordnance Survey. All rights reserved. 2009
A3a-g. Photos by Baca Architects (2008) Site photographs
A4. Ordnance Survey (2008) 1865 Ordnance Survey map showing former mill sites: © Crown copyright Ordnance Survey. All rights reserved. 2009
A5. Cities of Science (2008) Canalised section of the Wandle (online). www.citiesofscience.co.uk/go/London/ContentPlace_2830.html:
A6. Cities of Science (2008) Canalised section of the Wandle (online). www.citiesofscience.co.uk/go/London/ContentPlace_2830.html
A7. Baca Architects (2008) Topographical plan: Based on information provided by Ordnance Survey in 2008. © Crown copyright Ordnance Survey. All rights reserved. 2009
A8. Wandle valley (2008) Wandle trail (online). www.wandlevalleyfestival.org.uk/map.html
A9. LDA Design (2008) Local Landscape features
A10. EA (2008) Flood Risk Map (online). www.environment-agency.gov.uk © Environment Agency copyright 2009. All rights reserved
A11. Halcrow Group Ltd (2008) Flood hydrograph
A12. Halcrow Group Ltd (2008) Site 1 flood risk map current: Based on draft SFRA level 1 information provided by the Environment Agency in 2008, © Environment Agency copyright 2009. All rights reserved, © Crown copyright Ordnance Survey. All rights reserved. 2009
A13. Halcrow Group Ltd (2008) Site 1 flood risk map future: Based on draft SFRA level 1 information provided by the Environment Agency in 2008, © Environment Agency copyright 2009. All rights reserved, © Crown copyright Ordnance Survey. All rights reserved. 2009
A14. Baca Architects (2008) Outdoor Play areas
A15. Baca Architects (2008) Energy demand mix
A16. Baca Architects (2008) Spatial requirements for rainwater, attenuation, harvesting
A17. Baca Architects (2008) Plan options for the Felnex site
A18. Baca Architects (2008) Concept plan
A19. Baca Architects (2008) Conceptual Masterplan for Hackbridge Felnex and rail sites
A20. Baca Architects (2008) Energy technology contribution to total demand
A21. Baca Architects (2008) Space requirements for energy technologies
A22. Young Germany (2008) Solar panels at Vauban, Germany (Online). www.young-germany.de
A23. Baca Architects (2008) Locations of renewable energy infrastructure
A24. Civil Engineering 150, November 2002: Modern waterwheel, Bavaria
A25. Baca Architects (2008) Rainwater harvesting areas
A26. Baca Architects (2008) Development and flood risk
A27. Baca Architects (2008) Potential sources of flooding
A28. Photo by LDA Design (2008) Gravel swale
A29. Baca Architects (2008) Sustainable drainage detention basins, rain gardens and swales
A30. Baca Architects (2008) Drainage paths
A31. Baca Architects (2008) Safe havens, evacuation routes and flood stores
A32. Baca Architects (2008) Strategic flood measures
A33. Baca Architects (2008) Areas for outdoor play
A34. Baca Architects (2008) Dry pond and swales
A35. Baca Architects (2008) Local areas for play combined with rain gardens
A36a. Photo by LDA Design (2008) Mill Creek Canyon, USA
A36b. Photo by LDA Design (2008) Swale in Malmo, Sweden
A36c. Photo by LDA Design (2008) Eco Pond
A36d. Photo by LDA Design (2008) Preston Baggot, shallow lake
A37. Baca Architects (2008) Site Plan
A38a. Baca Architects (2008) Section through the River Wandle
A38b. Baca Architects (2008) Section through the Village Blue/Green
A38c. Baca Architects (2008) Section through a courtyard rain garden
A39a. Baca Architects (2009) LifE Handbook: View from the rain garden courtyard, looking towards the distant river, and village blue/green: IHS BRE Press
A39b. Baca Architects (2008) During a 1 in 20 year storm the rain gardens and swales retain rainwater, reducing run off overland and into the sewers.
A39c. Baca Architects (2008) During a 1 in 20 year storm the rain gardens and swales retain rainwater, reducing run off overland and into the sewers.
A40. Cyril Sweett (2008) Cost comparison between base and LifE option
A41. Cyril Sweett (2008) Cost between integrated and non-integrated approach
A42. Baca Architects (2008) Space demands for amenity, energy and water on the Felnex and rail site in Hackbridge

APPENDIX B

B1. Baca Architects (2008) Location Plan
B2. Baca Architects (2008) The site areas: © Crown copyright Ordnance Survey. All rights reserved. 2009
B3. Baca Architects (2008) Key to photographs: © Crown copyright Ordnance Survey. All rights reserved. 2009
B3a-f Photos by Baca Architects (2009) Site photographs
B4. Ordnance Survey (2008) 1898 Ordnance Survey map:
B5. Ordnance Survey (2008) 1936-1952 Ordnance Survey map:
B6. DEFRA (2008) Topographic Map: © Crown copyright 2009
B7 EA (2008) Historical Flood extents: © Environment Agency copyright 2009. All rights reserved.
B8. English Nature (2008) Nene washes (online). www.english-nature.org.uk © Natural England copyright 2009
B9. Bourne (2008) The Car Dyke (online): homepages.which.net/~rex/bourne/cardyke.htm:
B10. LDA Design (2008): Landscape character of the Nene corridor: © Crown copyright Ordnance Survey. All rights reserved. 2009
B11. LDA Design (2008): Landscape character of the Nene through Peterborough: © Crown copyright Ordnance Survey. All rights reserved. 2009
B12. EA (2008) Flood risk map (online). www.environment-agency.gov.uk
B13. Draft Catchment Flood Management Plan (unpublished) Accretion of flows
B14. EA (2008) Flood risk map current: © Environment Agency copyright 2009. All rights reserved

Image Credits

B15. Halcrow Group Ltd (2008) Flood hydrograph for 1 in 100 year event
B16. Halcrow Group Ltd (2008) Flood map, future: Based on CFMP modelling information provided by the Environment Agency in 2008: © Environment Agency copyright 2009. All rights reserved
B17. Halcrow Group Ltd (2008) Flood map, future: extrapolated from CFMP levels and not verified by the EA
B19. Baca Architects (2008) Spatial requirements for open play
B20. Baca Architects (2008) Energy demand mix
B21. Baca Architects (2008) Spatial requirements for rainwater, attenuation, harvesting
B22. Baca Architects (2008) Concept plan
B23. Baca Architects (2008) Plan options
B24. Baca Architects (2008) Conceptual Master plan for Peterborough Fengate site
B25. Baca Architects (2008) Energy technology contribution to meet total supply
B26. Baca Architects (2008) Space requirements for energy technologies
B27. Ecotricity (2008) Scales of main proven wind turbine technologies: © Ecotricity copyright 2009. All rights reserved
B28. Baca Architects (2008) Peterborough energy corridor
B29. Baca Architects (2008) Plan study of the spacing required for wind turbines
B30. Baca Architects (2008) Rainwater harvesting areas
B31. Baca Architects (2008) Development and flood risk
B32. Baca Architects (2008) Potential sources of flooding
B33. Baca Architects (2008) Site drainage
B34. Baca Architects (2008) Rainwater attenuation
B35. Baca Architects (2008) Terraced access routes
B36. Baca Architects (2008) Safe havens and exit routes
B37. Baca Architects (2008) Rainwater attenuation
B38. Baca Architects (2008) Sections
B39. Baca Architects (2008) Areas for outdoor play
B40. Photo by LDA Design (2008) Pollarded trees, characteristic of fenland landscapes help support biodiversity
B41. Photo by Baca Architects (2008) Water meadows within the centre of Cambridge
B42. Baca Architects (2008) Section through a stream corridor
B43. LDA Design (2008) Plan of stream corridor
B44a. Photo by Baca Architects (2008) Dutch waterside development
B44b. Photo by Baca Architects (2008) Dutch waterside development
B44c. Photo by Jeffrey Friedl (2007) Kamo River in Kyoto: http://regex.info/blog
B44d. Photo by Baca Architects (2008) Cambridge
B45. Baca Architects: Site Plan
B46a. Baca Architects (2008) Section through the River Nene
B46b. Baca Architects (2008) Section through the River Nene¬
B46c. Baca Architects (2008) Section through home zone and stream corridor
B46d. Baca Architects (2008) Section through rain corridor
B47a. Baca Architects (2009) LifE Handbook: View showing linear parkland drainage corridor and raised access to Home Zones: IHS BRE Press
B47b. Baca Architects (2008) During a 1 in 20 year storm the rain gardens and swales retain rainwater, reducing run off overland and into the sewers.
B47c. Baca Architects (2008) During a 1 in 100 year storm the rain gardens and swales retain large quantities of rainwater. The river is allowed to expand into predetermined areas of the village green and village blue.
B48. Cyril Sweett (2008) Cost between integrated and non-integrated approach
B49. Cyril Sweett (2008) Cost comparison between base and LifE option
B50. Baca Architects (2008) Space demands for amenity, energy and water on the riverside sites in Peterborough

APPENDIX C

C1. Baca Architects (2008) Location Plan
C2. Baca Architects (2008) The site: © Crown copyright Ordnance Survey. All rights reserved. 2009
C3. Baca Architects (2008) Key to photographs: © Crown copyright Ordnance Survey. All rights reserved. 2009
C3a-e. Photos by Baca Architects (2008) Site Photographs
C4a. West Sussex District Council (2008) 1587 Armada map
C4b. West Sussex District Council (2008) From an early plan of Arundel Haven 1736.
C4c. West Sussex District Council (2008)1698 South Coast Harbours Report, Edward Dummers and Thomas Wiltshaw
C4d. Ordnance Survey (2008) 1879 Map. © Crown copyright Ordnance Survey. All rights reserved. 2009
C5. Photo by Baca Architects (2008) Low-lying farmland to the north of the coast
C6. Photo by Baca Architects (2008) The dredging works to the east bank of the River Arun
C7. Royal Haskoning (2008) Topographic map, Tidal River Arun, Initial Geomorphological Assessment, Draft Report 2, Royal Haskoning for Environment Agency, West Sussex, 23 April 2007
C8 EA (2008) River catchment system (online) www.environment-agency.gov.uk
C9. EA (2008) Contour map (online) www.environment-agency.gov.uk
C10. Baca Architects (2008) View of Arundel from the River Arun
C11. Baca Architects (2008) Climping Beach
C12. LDA Design (2008) Regional landscape character: based on information from Ordnance Survey. © Crown copyright Ordnance Survey. All rights reserved. 2009
C13. LDA Design (2008) Local landscape character: based on information from Ordnance Survey. © Crown copyright Ordnance Survey. All rights reserved. 2009
C14. EA (2008) Flood risk map (online) www.environment-agency.gov.uk:
C15. EA (2008) Flood risk map (online) www.environment-agency.gov.uk:
C16. Halcrow Group Ltd (2008) Flood map, with 1 m sea level rise (100 years based on Defra guidance, October 2006): © Environment Agency copyright 2009. All rights reserved. © Crown copyright Ordnance Survey. All rights reserved. 2009
C17. Halcrow Group Ltd (2008) Flood map, with 2 m sea level rise: © Environment Agency copyright 2009. All rights reserved. © Crown copyright Ordnance Survey. All rights reserved. 2009
C19. Baca Architects (2008) Spatial requirements for open play
C20. Baca Architects (2008) Energy demand mix
C21. Baca Architects (2008) Spatial requirements for rainwater, attenuation, harvesting
C22. Baca Architects (2008) Concept plan
C23. Baca Architects (2008) Plan options for harbour site
C24. Baca Architects (2008) Conceptual Masterplan for Littlehampton
C25. Fulcrum Consulting (2008) Energy technology contribution to total demand
C26. Baca Architects (2008) Space requirements for option 2
C27a. Ecotricity (2008) 85 m high 2 MW wind turbines at Green Park Reading (top) and Ford, Dagenham © Ecotricity copyright 2009. All rights reserved
C28. Baca Architects (2008) Location of wind turbines within the lower Arun site area.
C29. Baca Architects (2008) Location of tidal power stations within the harbour development
C30. Halcrow group Ltd (2008) Littlehampton UNA Wharf - approximate tidal curve and potential power generating periods
C31. Baca Architects (2008) Rainwater harvesting areas
C32a. Baca Architects (2008) Exploratory development plan 1 m Sea Level Rise
C32b. Baca Architects (2008) Exploratory development plan 3 m Sea Level Rise
C32c. Baca Architects (2008) Exploratory development plan 6 m Sea Level Rise
C33. Halcrow group Ltd (2008) Flood storage areas
C34. Halcrow Group Ltd (2008) Flood map, current provided by the Environment Agency in 2008: © Environment Agency copyright 2009. All rights reserved
C35. Baca Architects (2008) Flood zones and development* Note flood zones after development are indicated
C36. Baca Architects (2008) Potential sources of flooding
C37. Baca Architects (2008) Safe havens, evacuation routes and flood stores
C38. Baca Architects (2008) Residential units with elevated access constructed above retail and workshops
C39. Baca Architects (2008) SUDS areas and drainage paths
C40. Halcrow Group Ltd (2008) Potential Habitat Creation Map: © Environment Agency copyright 2009. All rights reserved.
C41. Halcrow Group Ltd (2008) Potential Habitat with 1.045 m sea level rise: © Environment Agency copyright 2009. All rights reserved
C42. Baca Architects (2008) Potential saltmarsh plant species
C43. Baca Architects (2008) Landscape areas
C44. Photo by Dura Vermeer Group (2008) Flood resilient play area
C45a. Photo by LDA Design (2008) Moorings on the River Welland, Lincolnshire
C45b. Photo by Dura Vermeer Group (2008) Amphibious and floating waterside development in Maasbommel, the Netherlands
C45c. Baca Architects (2008) Dordrecht floodproof pilot project
C45d. Photo by Baca Architects (2008) Tilbury marshes, Thames gateway, Essex
C46. Baca Architects (2008) Site Plan
C47a. Baca Architects (2008) Section (vertical scale of landscape exaggerated)
C47b. Baca Architects (2008) Section through the River Arun
C47c. Baca Architects (2008) Section through salt marsh (vertical scale of landscape exaggerated)
C47d. Baca Architects (2008) Section through tidal lagoon (vertical scale of landscape exaggerated)
C48a. Baca Architects (2009) LifE Handbook: View, showing linear development, boulder swales, promenade and distant estuarine lagoon: IHS BRE Press
C48b. Baca Architects (2008) During a 1 in 20 year storm rainwater is retained in rain courtyards, reducing run off overland and into the sewers.
C48c. Baca Architects (2008) During a 1 in 200 year storm peak water is diverted into the distant lagoon, reducing river levels. The courtyards retain rainwater that cannot drain into the river and in the unlikely event of a river wall breach, high pathways allow safe access to property.
C49. Cyril Sweett (2008) Cost between integrated and non-integrated approach
C50. Cyril Sweett (2008) Cost comparison between base and LifE option
C51. Baca Architects (2008) Total space demands for the Climping and Harbour Sites at Littlehampton.

APPENDIX D

D1. Baca Architects (2009): Site 1 flood risk map current (2009) based on SFRA level 1 © Environment Agency copyright 2009. All rights reserved. © Crown copyright Ordnance Survey. All rights reserved. 2009
D2. Environment Agency (2009): Site 2 flood risk map current © Environment Agency copyright 2009. All rights reserved. © Crown copyright Ordnance Survey. All rights reserved. 2009

Glossary

AOD Above Ordnance Datum - measure of height above standard datum level

ATES Aquifer Thermal Energy Storage uses the temperature difference between an underground temperature store (in this case an aquifer) and a building above ground to heat and cool the building.

Biomass is the biodegradable fraction of products, waste and residues from agriculture (including plant and animal substances), forestry and related industries, as well as the biodegradable fraction of industrial and municipal waste.[28]

Brownfield Land (common name for Previously-developed land) 'Previously-developed land is that which is or was occupied by a permanent structure, including the curtilage of the developed land and any associated fixed surface infrastructure.'[26]

CFMP Catchment Flood Management Plan

Climate Change - is any long-term significant change in the "average weather" that a given region experiences.

Code Code for Sustainable Homes. The Code is intended as a national standard to guide industry in the design and construction of sustainable homes. The Code measures the sustainability of a home against nine design categories: energy, pollution, water, health and well-being, materials, management, surface water run off, ecology and waste.

CLG Communities and Local Government sets policy on local government, housing, urban regeneration, planning and fire and rescue.

Defra Department for Environment Food and Rural Affairs. Defra has overall policy responsibility for flood and coastal erosion risk in England.

Ecological Flood Mitigation - flood mitigation techniques based on use of natural systems or mechanisms.

EP English Partnerships, see HCA

EA The Environment Agency are the UK government agency concerned mainly with rivers, flooding, and pollution. Responsible for mapping flood risk and managing: flood defences, river channels and shorelines in England and Wales.

FZ Flood Zone is an area of land with a designated probability of flooding.

Floodproof - designed to withstand a flood

Fluvial Flooding - flooding from rivers

FRA Flood Risk Assessment

FCERM Flood and Coastal Erosion Risk Management

GIFA Gross Internal Floor Area is the area measure of all floors and circulation spaces within a building.

Green Infrastructure - refers to a network of open spaces and nature corridors that can be used to manage flood risk, provide wildlife corridors and amenity space instead of civil engineered infrastructure.

GSHP Ground Source Heat Pumps exchange heat between the ground and a building to heat and cool the building.

HCA Homes and Communities Agency

ITS Interseasonal Thermal Storage uses an underground storage area to store heat gained from the sun in the summer. A heat pump then transfers this heat from the store to a building in the winter time when it is needed.

LAP Local Area for Play

LEAP Local Equipped Area for Play

NEAP Neighbourhood Equipped area of Play

MUGA Multi-use Gaming Area

MHWN Mean High Water Neap tide level

MHWS Mean High Water Springs tide level

MLWN Mean Low Water Neap tide level

MLWS Mean Low Water Springs tide level

MSW Making Space for Water - see pages 4-5

ODN Ordnance Datum Newlyn (ODN) - level above or below standard datum level

Pluvial Flooding - flooding from rain

PPS Planning Policy Statement's replace the former Planning Policy Guidance with national policy on spatial planning.

PPG Planning Policy Guidance

PV Photo Voltaic – A solar PV is a large area panel that converts solar energy into electricity by the photovoltaic effect.

Renewable Energy – Energy generated from naturally renewable resources, such as wind, sun and water. This also included crops such as willow

RO The Renewables Obligation requires licensed electricity suppliers to source a specific and annually increasing percentage of the electricity they supply from renewable sources.

Resilience - Designed to withstand flooding.

Return Period - the period in which an event, such as a high spring tide, is expected to recur.

Shadow Flicker - intermittent shadow cast by wind turbines.

SSSI Site of Special Scientific Interest

SFRA Strategic Flood Risk Assessment

SUDS Sustainable Drainage Systems are designed to manage rainfall on development in a way that mimics natural drainage, reducing or without the need for piped underground drainage.

Tidal Flooding - flooding from the sea or tidal parts of the river/estuary.

URC Urban Regeneration Company

Zero Carbon - Providing all energy needs from renewable resources on site, such as wind, tidal and solar power. See page 188 for a detailed definition.

Glossary

Tewkesbury 2007

Flooding from rivers (fluvial flooding)

Flooding can either develop gradually or rapidly according to how steeply the ground rises in the catchment and how fast water runs off into surface watercourses. In a large, relatively flat catchment, flood levels will rise slowly and land may remain flooded for several days.

Duration: days *Risk: medium to high* *Warning time: short to long*

Leigh-on-Sea 1953

Flooding from the sea (tidal flooding)

Flooding to low-lying land from the sea and tidal estuaries is caused by storm surges and high tides. Where flood defences exist, they can be over topped or breached during a severe storm. The onset of flooding from the sea can be extremely rapid. Deep, fast-flowing water can create an extreme hazard.

Duration: hours *Risk: high* *Warning time: medium*

Balham 2007

Flooding from land (surface water flooding)

Intense rainfall can run quickly off land and result in local (pluvial) flooding. This is more likely in areas with low permeability such as built up areas, where flooding can also be polluted by run off and sewage.

Duration: hours *Risk: medium* *Warning time: short*

Coventry 2004

Flooding from drains and sewers

In many towns foul and surface (rain) water are drained into combined sewers. Heavy rainfall can overwhelm the sewers causing sewage to spill onto the streets and into properties. Contaminants can cause health risk to people and pollution of water channels, damaging local ecology.

Duration: hours to days *Risk: medium* *Warning time: short to medium*

Hull 2007

Flooding from groundwater

Groundwater flooding is most likely to occur in low lying areas underlain by permeable rocks. Groundwater levels will rise during wetter months. In particularly wet weather the water may rise above the level of the ground, causing flooding that can last for weeks or months.

Duration: weeks to months *Risk: low* *Warning time: long*

Teesdale Reservoir (example only)

Flooding from infrastructure failure

Sources of flooding include reservoirs, canals, lakes and industrial processes, including mining, quarrying and sand or gravel extraction. Flooding may occur if excess water overwhelms the facility and/or if the retaining structure fails. Sudden failure can result in rapidly flowing, deep water, causing significant threat to life and property.

Duration: varies *Risk: medium to high* *Warning time: short*

Types of flooding (from PPS25, Annex C) and indicative characteristics.

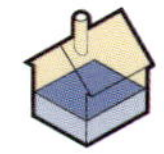

Resilient buildings

Buildings made to be resistant or resilient to flooding to reduce cost of damages and aid recovery.

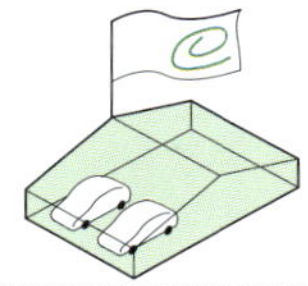

Car club

Vehicles are shared by members of a club. A typical car club has up to 20 members for each vehicle.

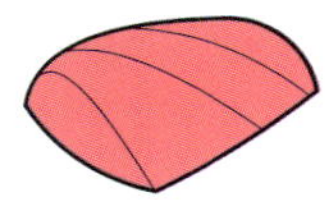

Safe havens

Buildings designed to a higher standard than those surrounding to provide secure refuge during a flood.

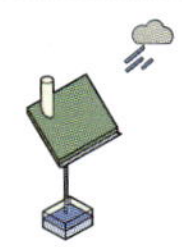

Green roof

A planted roof to a building that helps to soak up rainwater, slowing the flow down to the ground.

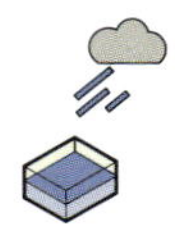

Rainwater harvesting

Collecting, storing and reusing rainwater that falls on and around buildings.

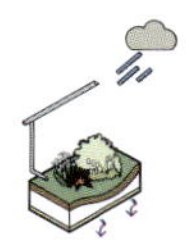

Rain garden

A shallow, planted depression used to store excessive rainwater before it soaks away.

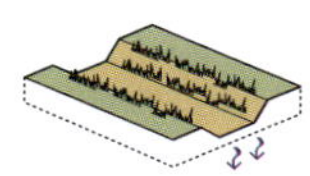

Swales

Shallow channels that are designed to convey, infiltrate, store and treat run off rainwater.

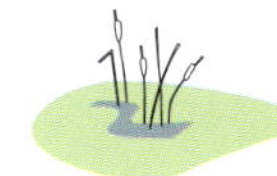

Wetlands

Land regularly flooded such as salt marshes, mudflats, swamps, bogs and fens that can support biodiversity.

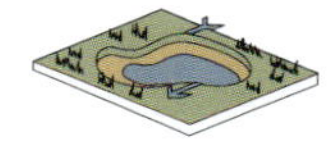

Detention ponds

Standing areas of water or dry depressions used to store excess rain water before directing it into the water course.

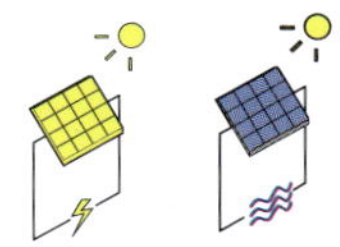

Solar panels

Solar photovoltaic cells (PVs) convert the sun's energy into electricity. Solar thermal panels warm water.

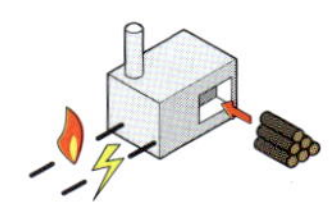

Biomass CHP

Small powerstations that provide Combined Heat and Power from burning fuel such as wood chip.

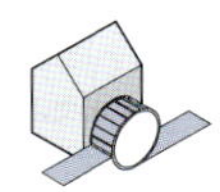

Micro-hydro power

Water flow with a fall of at least 1.5 m can be used to turn a water wheel or 'Archimedes screw' to generate power.

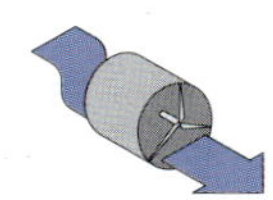

Tidal power

Use of tidal stream or tidal range to drive turbines and generate power. Large areas for retaining water are required.

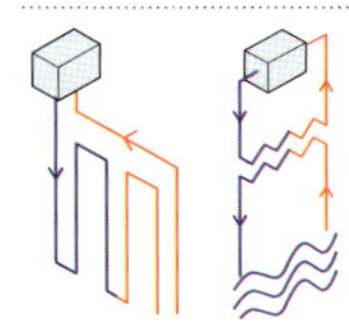

GSHP and ATES

Ground source heat pumps (GSHP) and aquifer thermal energy storage (ATES) exchange heat energy with the ground and underground water respectively.

The LifE Handbook

The LifE Handbook

The LifE handbook provides an introduction to flooding and development. The purpose of this handbook is to raise awareness, stimulate thought and aid delivery of more sustainable development. It is to be used by decision makers, designers and developers before and during the early stages of design, in conjunction with policy documents, such as Planning Policy Statement 25 (PPS25).

It contains useful references to other sources of information and further reading. The illustrated examples and case studies provide a brief overview of options that may be transferrable to other sites. However, each site has unique issues and problems which must be assessed by an expert team.

Contents

- The LifE approach
- Flood risk & development
- Types of flooding and location
- Designing for climate change
- Integrated design approach
- Upper catchment - let rain slow (Case study - Hackbridge)
- Middle catchment - let rivers flow (Case study - Hackbridge)
- Lower catchment - let tides go (Case study - Littlehampton)
- Cost analysis
- LifE lessons - make space for water, energy and play
- Illustrated glossary
- References
- Index

IHS BRE Press, Ref: EP 97, ISBN: 978-1-84806-088-3, 28 pages

The handbook is available from www.brebookshop.com

Comments

'The Homes and Communities Agency welcomes this guidance, as it takes what can be difficult to grasp issues and places them in an easily accessible context.'

Owen Peat, Project Manager – Environment National Consultancy Unit, Homes and Communities Agency (HCA)

'The LifE Handbook is a good introduction to flood risk and the implications for development; and even the experienced flood-risk practitioner will learn something from the LifE approach.'

Barrie Neaves, Growth Manager (Kent & East Sussex), Environment Agency

'It's a model of distilled expertise and a great example of an architect-led initiative with benefits for all'

Hugh Pearman, Editor of RIBA Journal

'The LifE handbook presents information about flood risk management in an informative and clear way, providing the target audience with focused and easy to use guidance. It is a very useful additional to the available guidance.'

Paul Shaffer, Associate, CIRIA.

'The handbook is an excellent and approachable document that is clear and thought provoking.'

Jonathan Harbottle, Director, Green Initiatives Development

'The LifE project encourages much-needed integrated thinking about the complex challenges facing urban development in the 21st century. The guidance is full of creative solutions and thought-provoking design approaches.'

Anna Penning-Rowsell, Enabling advisor, CABE